Optoelectronics: A Formula Handbook

N.B. Singh

DEDICATION

To Nature,

I dedicate this book to you, the source of all life. You are my inspiration, my teacher, and my friend.

Thank you for teaching me about the beauty of the world around me. Thank you for showing me the power of the natural world. Thank you for giving me a sense of peace and tranquillity.

I promise to do my part to protect you and your many wonders. I will teach my children about the importance of conservation and sustainability. I will work to make the world a better place for all living things.

Thank you for everything, Nature.

With love,

N.B Singh

Contents

Preface

In the rapidly advancing field of optoelectronics, the need for a concise and comprehensive reference guide has become increasingly evident. This formula handbook aims to fill that gap by providing a collection of essential formulas, equations, and practical examples related to optoelectronic devices and systems.

Scope and Purpose

The handbook covers a wide range of topics, including the fundamental principles of optics, semiconductor physics, and various optoelectronic devices. It is designed to be a quick and accessible resource for students, researchers, and professionals working in the field of optoelectronics.

Organization of the Handbook

The content is organized into several chapters, each focusing on a specific aspect of optoelectronics. The formulas presented are accompanied by explanations and, where applicable, real-world examples to facilitate a better understanding of the concepts.

How to Use This Handbook

Whether you are a student looking for quick references during studies, a researcher seeking formulas for experimental setups, or an engineer involved in the design of optoelectronic systems, this handbook is structured to meet your needs. Each chapter is self-contained, allowing readers to easily locate and utilize the relevant information.

Thank you for choosing "Optoelectronics: A Formula Handbook." I hope it proves to be a valuable resource in your journey through the fascinating world of optoelectronics.

Chapter 1

Introduction

Optoelectronics is a fascinating field that encompasses the study and application of devices that interact with light and electricity. In this introductory chapter, we will explore the fundamental concepts and key principles that form the basis of optoelectronics.

1.1 Basic Principles

The foundation of optoelectronics lies in the understanding of the interaction between light and semiconductor materials. The behavior of photons and electrons in these materials is crucial for the design and operation of optoelectronic devices.

1.1.1 Wave-Particle Duality

One of the fundamental principles in optoelectronics is the wave-particle duality of light. According to quantum mechanics, light exhibits both wave-like and particle-like characteristics. This duality is described by the de Broglie wavelength formula:

$$\lambda = \frac{h}{p} \tag{1.1}$$

where λ is the wavelength, h is Planck's constant, and p is the momentum of the particle.

1.1.2 Semiconductor Physics

The behavior of electrons in semiconductor materials is crucial for optoelectronic device operation. The carrier concentration in a semiconductor is determined by the Fermi-Dirac distribution:

$$f(E) = \frac{1}{1 + e^{(E - E_f)/(kT)}} \tag{1.2}$$

where E_f is the Fermi energy, k is the Boltzmann constant, and T is the temperature.

1.2 Optoelectronic Devices

Optoelectronic devices leverage the principles of light-matter interaction to perform various functions. Let's explore some key devices and their applications.

1.2.1 Light Emitting Diodes (LEDs)

LEDs are semiconductor devices that emit light when forward-biased. The radiative recombination in the semiconductor creates photons, resulting in light emission. The relationship between the energy of emitted photons and the bandgap energy of the material is given by:

$$E = \frac{hc}{\lambda} = E_{\text{gap}} \tag{1.3}$$

where E is the energy, h is Planck's constant, c is the speed of light, λ is the wavelength, and E_{gap} is the bandgap energy.

Example: LED Application

Consider an LED with a bandgap energy of 2 eV. Calculate the wavelength of the emitted light.

$$E_{\text{gap}} = 2\,\text{eV} \tag{1.4}$$

$$\lambda = \frac{hc}{E_{\text{gap}}}$$

$$= \frac{(6.626 \times 10^{-34}\,\text{J s} \times 3 \times 10^{8}\,\text{m/s})}{2\,\text{eV} \times 1.602 \times 10^{-19}\,\text{J/eV}}$$

$$\approx 620\,\text{nm} \tag{1.5}$$

The emitted light has a wavelength of approximately 620 nm.

1.2.2 Photodiodes

Photodiodes are semiconductor devices that convert light into electrical current. The photocurrent generated is proportional to the incident light intensity. The relationship between the photocurrent and incident light power is given by:

$$I_{\text{ph}} = \frac{P_{\text{in}} \cdot q}{h\nu} \tag{1.6}$$

where I_{ph} is the photocurrent, P_{in} is the incident light power, q is the elementary charge, h is Planck's constant, and ν is the frequency of the light.

Example: Photodiode Response

A photodiode receives an optical power of 10 mW at a wavelength of 850 nm. Calculate the photocurrent assuming ideal conditions.

$$P_{\text{in}} = 10\,\text{mW} = 0.01\,\text{W}$$

$$\nu = \frac{c}{\lambda}$$

$$= \frac{3 \times 10^8\,\text{m/s}}{850 \times 10^{-9}\,\text{m}}$$

$$\approx 3.53 \times 10^{14}\,\text{Hz}$$

$$I_{\text{ph}} = \frac{0.01\,\text{W} \cdot 1.602 \times 10^{-19}\,\text{C}}{6.626 \times 10^{-34}\,\text{J s} \times 3.53 \times 10^{14}\,\text{Hz}}$$

$$\approx 2.39 \times 10^{-9}\,\text{A} \tag{1.7}$$

The photocurrent is approximately 2.39×10^{-9} A.

1.3 Conclusion

This introductory chapter has provided a glimpse into the foundational principles of optoelectronics and introduced key devices such as LEDs and photodiodes. In the following chapters, we will delve deeper into specific topics, exploring advanced concepts and practical applications.

Chapter 2

Basic Optoelectronic Components

In this chapter, we delve into the fundamental components of optoelectronics, starting with Light Emitting Diodes (LEDs). LEDs are semiconductor devices that emit light when current flows through them. Understanding the underlying principles and characteristics of LEDs is essential for their proper integration into electronic systems.

2.1   Light Emitting Diodes (LEDs)

LEDs are crucial components in optoelectronics, widely used for lighting, displays, and indicators. The operation of an LED is based on the phenomenon of electroluminescence, where light is emitted due to the recombination of charge carriers in a semiconductor material. The basic equation governing the energy of emitted photons is given by:

$$E = \frac{hc}{\lambda} = E_{\text{gap}} \tag{2.1}$$

where E is the energy, h is Planck's constant, c is the speed of light, λ is the wavelength, and E_{gap} is the bandgap energy of the semiconductor material.

The efficiency of an LED can be described by the external quantum efficiency (EQE), representing the ratio of emitted photons to injected electrons:

$$\text{EQE} = \frac{\text{Number of Photons Emitted}}{\text{Number of Electrons Injected}} \tag{2.2}$$

2.1.1   Example: Calculating Wavelength

Consider an LED with a bandgap energy (E_{gap}) of 2 eV. Calculate the corresponding wavelength (λ) of the emitted light using the energy-wavelength relationship.

$$E_{\text{gap}} = 2\,\text{eV} \tag{2.3}$$

$$\lambda = \frac{hc}{E_{\text{gap}}}$$

$$= \frac{(6.626 \times 10^{-34}\,\text{J s} \times 3 \times 10^{8}\,\text{m/s})}{2\,\text{eV} \times 1.602 \times 10^{-19}\,\text{J/eV}}$$

$$\approx 620\,\text{nm} \tag{2.4}$$

The emitted light has a wavelength of approximately 620 nm.

2.1.2   LED Characteristics

LEDs exhibit various characteristics, including forward voltage (V_f), forward current (I_f), and luminous intensity (I_v). These characteristics are crucial for designing and operating LED circuits.

The relationship between forward voltage and forward current in an LED can be described by the Shockley diode equation:

$$I_f = I_s \left(e^{\frac{V_f}{nV_t}} - 1 \right) \tag{2.5}$$

where I_s is the reverse saturation current, V_t is the thermal voltage, and n is the ideality factor.

2.1.3 Example: LED Characteristics

Consider an LED with a forward voltage of 2 V, a forward current of 20 mA, and an ideality factor of 1.2. Calculate the reverse saturation current (I_s).

$$V_f = 2\,\text{V}$$

$$I_f = 20\,\text{mA} = 0.02\,\text{A}$$

$$n = 1.2$$

$$I_s = \frac{I_f}{e^{\frac{V_f}{nV_t}} - 1}$$

$$\approx 1.43 \times 10^{-12}\,\text{A} \tag{2.6}$$

The reverse saturation current is approximately 1.43×10^{-12} A.

2.1.4 LED Efficiency

The efficiency of an LED is crucial for its practical applications. The power efficiency (η) of an LED is given by the ratio of optical power output (P_{out}) to electrical power input (P_{in}):

$$\eta = \frac{P_{\text{out}}}{P_{\text{in}}} \tag{2.7}$$

2.1.5 Example: LED Efficiency

Consider an LED with an optical power output of 10 mW and an electrical power input of 20 mW. Calculate the power efficiency of the LED.

$$P_{\text{out}} = 10\,\text{mW} = 0.01\,\text{W}$$

$$P_{\text{in}} = 20\,\text{mW} = 0.02\,\text{W}$$

$$\eta = \frac{0.01\,\text{W}}{0.02\,\text{W}}$$

$$= 0.5 \tag{2.8}$$

The power efficiency of the LED is 0.5 or 50

In summary, Light Emitting Diodes (LEDs) play a crucial role in optoelectronics, emitting light efficiently based on electroluminescence. Understanding the characteristics and efficiency of LEDs is essential for their successful integration into electronic systems.

2.2 Photodiodes

Photodiodes are crucial components for detecting and converting optical signals into electrical signals. The operation of a photodiode relies on the generation of photocurrent when exposed to light. The relationship between the photocurrent (I_{ph}) and incident light power (P_{in}) can be expressed by the formula:

$$I_{\mathrm{ph}} = \frac{P_{\mathrm{in}} \cdot q}{h\nu} \tag{2.9}$$

where I_{ph} is the photocurrent, P_{in} is the incident light power, q is the elementary charge, h is Planck's constant, and ν is the frequency of the incident light.

2.2.1 Example: Calculating Photocurrent

Consider a photodiode exposed to optical power (P_{in}) of 5 mW at a wavelength of 700 nm. Calculate the resulting photocurrent (I_{ph}) assuming ideal conditions.

$$P_{\text{in}} = 5\,\text{mW} = 0.005\,\text{W}$$

$$\lambda = 700\,\text{nm}$$

$$\nu = \frac{c}{\lambda}$$

$$= \frac{3 \times 10^8\,\text{m/s}}{700 \times 10^{-9}\,\text{m}}$$

$$\approx 4.29 \times 10^{14}\,\text{Hz}$$

$$I_{\text{ph}} = \frac{0.005\,\text{W} \cdot 1.602 \times 10^{-19}\,\text{C}}{6.626 \times 10^{-34}\,\text{J s} \times 4.29 \times 10^{14}\,\text{Hz}}$$

$$\approx 1.86 \times 10^{-9}\,\text{A} \tag{2.10}$$

The resulting photocurrent is approximately 1.86×10^{-9} A.

2.2.2 Photodiode Characteristics

The performance of photodiodes is characterized by various parameters, including responsivity (R), quantum efficiency (QE), and bandwidth (BW). These parameters are critical for evaluating the suitability of a photodiode for specific applications.

The responsivity of a photodiode is defined as the ratio of the generated photocurrent to the incident optical power:

$$R = \frac{I_{\text{ph}}}{P_{\text{in}}} \tag{2.11}$$

The quantum efficiency represents the efficiency with which photons are converted into electron-hole pairs in the photodiode material:

$$QE = \frac{\text{Number of Electron-Hole Pairs Generated}}{\text{Number of Incident Photons}} \tag{2.12}$$

The bandwidth of a photodiode refers to its ability to respond to changes in incident light intensity. It is a crucial parameter for high-speed communication systems.

2.2.3   Example: Photodiode Characteristics

Consider a silicon photodiode with a responsivity of 0.9 A/W and a quantum efficiency of 85

$$R = 0.9\,\text{A/W}$$

$$QE = 0.85$$

$$P_{\text{in}} = 2\,\text{mW} = 0.002\,\text{W}$$

$$I_{\text{ph}} = R \cdot P_{\text{in}}$$

$$= 0.9\,\text{A/W} \cdot 0.002\,\text{W}$$

$$= 0.0018\,\text{A}$$

$$I_{\text{ph, QE}} = I_{\text{ph}} \cdot QE$$

$$= 0.0018\,\text{A} \cdot 0.85$$

$$= 0.00153\,\text{A} \tag{2.13}$$

The resulting photocurrent, considering quantum efficiency, is approximately 0.00153 A.

In summary, photodiodes play a crucial role in light detection and communication systems. Understanding their characteristics, including responsivity, quantum efficiency, and bandwidth, is essential for selecting the appropriate photodiode for a given application.

2.3   Phototransistors

Phototransistors are three-terminal devices that exhibit the characteristics of a transistor while responding to incident light. The basic structure includes a photodiode integrated with a transistor. The photocurrent generated by the photodiode controls the base current of the transistor, leading to amplified collector current.

The relationship between the collector current (I_c), base current (I_b), and amplification factor (h_{fe}) of the phototransistor can be expressed by the following formula:

$$I_c = h_{fe} \cdot I_b \tag{2.14}$$

where h_{fe} is the common emitter current transfer ratio.

2.3.1 Example: Phototransistor Amplification

Consider a silicon phototransistor with a common emitter current transfer ratio of 100. If the base current (I_b) is 10 ηA, calculate the resulting collector current (I_c).

$$
\begin{aligned}
h_{fe} &= 100 \\
I_b &= 10\,\mu\text{A} = 10^{-5}\,\text{A} \\
I_c &= h_{fe} \cdot I_b \\
&= 100 \cdot 10^{-5} \\
&= 0.001\,\text{A} = 1\,\text{mA}
\end{aligned}
\tag{2.15}
$$

The resulting collector current is 1 mA.

2.3.2 Phototransistor Characteristics

Phototransistors offer advantages such as high sensitivity, fast response times, and ease of integration into electronic circuits. The sensitivity of a phototransistor is characterized by the responsivity (R_{PT}), representing the ratio of the change in collector current to the change in incident light power:

$$R_{\text{PT}} = \frac{\Delta I_c}{\Delta P_{\text{in}}} \tag{2.16}$$

The bandwidth (BW_{PT}) of a phototransistor is crucial for applications requiring rapid changes in light intensity. It represents the frequency range over which the phototransistor can effectively respond.

2.3.3　Example: Responsivity and Bandwidth

For a specific silicon phototransistor, the responsivity is measured to be $0.8\,\mathrm{A/W}$, and the bandwidth is $100\,\mathrm{kHz}$. Calculate the change in collector current for a change in incident light power of $1\,\mathrm{mW}$.

$$R_{\mathrm{PT}} = 0.8\,\mathrm{A/W}$$

$$\Delta P_{\mathrm{in}} = 1\,\mathrm{mW} = 0.001\,\mathrm{W}$$

$$\Delta I_c = R_{\mathrm{PT}} \cdot \Delta P_{\mathrm{in}}$$

$$= 0.8\,\mathrm{A/W} \cdot 0.001\,\mathrm{W}$$

$$= 0.0008\,\mathrm{A} = 0.8\,\mathrm{mA} \tag{2.17}$$

The change in collector current is $0.8\,\mathrm{mA}$ for a $1\,\mathrm{mW}$ change in incident light power.

In summary, phototransistors play a vital role in optoelectronic systems by providing amplified light detection. Understanding their operation, characteristics, and parameters like current transfer ratio, responsivity, and bandwidth is crucial for designing and implementing effective optoelectronic circuits.

Chapter 3

Optical Sensors

3.1 Position Sensors

Optical position sensors are devices that detect and measure the position of an object with respect to a reference point. One common type of optical position sensor is the encoder, which converts the angular or linear position of an object into an electrical signal. The relationship between the position (P), the number of pulses (N) generated by the encoder, and the resolution (R) can be expressed as follows:

$$P = N \times R \tag{3.1}$$

where P is the position, N is the number of pulses, and R is the resolution.

3.1.1 Example: Encoder Resolution

Consider an optical encoder with a resolution of 1000 pulses per revolution. If the encoder generates 5000 pulses during an operation, calculate the corresponding angular position.

$$N = 5000$$

$$R = 1000$$

$$P = N \times R$$

$$= 5000 \times 1000$$

$$= 5 \times 10^6 \text{ units} \tag{3.2}$$

The angular position is 5×10^6 units.

3.1.2 Absolute vs. Incremental Encoders

Position sensors can be broadly classified into absolute and incremental encoders. Absolute encoders provide the absolute position of an object within a
single revolution, while incremental encoders generate pulses corresponding to
the changes in position.

The resolution of an incremental encoder is crucial for determining the ac-
curacy of position measurements. It is defined as the number of pulses per unit
of movement (e.g., pulses per millimeter for linear encoders).

3.1.3 Example: Linear Encoder Resolution

Consider a linear encoder with a resolution of 2000 pulses per millimeter. If
the encoder generates 8000 pulses during the linear motion, calculate the corresponding distance traveled.

$$N = 8000$$

$$R = 2000 \, \text{pulses/mm}$$

$$P = N/R$$

$$= \frac{8000}{2000} \, \text{mm}$$

$$= 4 \, \text{mm} \tag{3.3}$$

The distance traveled is 4 mm.

3.2 Optical Sensors in Robotics

Position sensors are integral components in robotic systems, providing real-time feedback on the position and orientation of robotic arms or end-effectors. The information obtained from these sensors is crucial for precise control and manipulation of robotic movements.

3.2.1 Robot Arm Control

In a robotic arm, optical encoders are often used to monitor the joint angles. The relationship between the joint angle (θ), the number of encoder pulses (N), and the resolution (R) can be expressed as:

$$\theta = \frac{N}{R} \tag{3.4}$$

This information allows the robotic controller to accurately determine the position of each joint.

3.2.2 Example: Robotic Arm Position

Consider a robotic arm joint with an optical encoder having a resolution of 500 pulses per degree. If the encoder generates 2500 pulses, calculate the corresponding joint angle.

$$N = 2500$$

$$R = 500 \, \text{pulses/degree}$$

$$\theta = \frac{N}{R}$$

$$= \frac{2500}{500} \, \text{degrees}$$

$$= 5 \, \text{degrees} \tag{3.5}$$

The joint angle is 5 degrees.

Position sensors, especially optical sensors like encoders, play a crucial role in optoelectronics, enabling accurate position measurements in various applications. The formulas and examples presented in this section provide a foundation for understanding and utilizing optical position sensors in practical scenarios.

3.3 Proximity Sensors

Proximity sensors operate based on the principles of light reflection or transmission to detect the presence or absence of an object. One common type is the infrared (IR) proximity sensor. The relationship between the intensity of the reflected infrared light (I_r), the incident light intensity (I_i), and the distance (d) can be expressed using the inverse square law:

$$I_r = \frac{I_i}{\left(1 + \left(\frac{d}{D}\right)^2\right)} \tag{3.6}$$

where D is the effective detection range of the proximity sensor.

3.3.1 Example: IR Proximity Sensor

Consider an IR proximity sensor with an incident light intensity (I_i) of $10\,\mathrm{mW/cm}^2$ and an effective detection range (D) of $5\,\mathrm{cm}$. Calculate the intensity of the reflected light (I_r) at a distance (d) of $2\,\mathrm{cm}$.

$$I_i = 10\,\mathrm{mW/cm}^2$$

$$d = 2\,\mathrm{cm}$$

$$D = 5\,\mathrm{cm}$$

$$I_r = \frac{I_i}{\left(1 + \left(\frac{d}{D}\right)^2\right)}$$

$$= \frac{10}{\left(1 + \left(\frac{2}{5}\right)^2\right)}$$

$$\approx 6.25\,\mathrm{mW/cm}^2 \tag{3.7}$$

The reflected light intensity is approximately $6.25\,\text{mW/cm}^2$ at a distance of $2\,\text{cm}$.

3.3.2 Detection Range and Sensitivity

The detection range (D) of a proximity sensor is a critical parameter. It represents the maximum distance at which the sensor can reliably detect an object. The sensitivity (S) of a proximity sensor is defined as the change in output per unit change in distance:

$$S = \frac{\Delta\text{Output}}{\Delta d} \tag{3.8}$$

3.3.3 Example: Proximity Sensor Sensitivity

For a proximity sensor with a detection range (D) of $10\,\text{cm}$ and an output change of $4\,\text{V}$ over a distance change (Δd) of $2\,\text{cm}$, calculate the sensitivity (S).

$$D = 10\,\text{cm}$$
$$\Delta d = 2\,\text{cm}$$
$$\Delta\text{Output} = 4\,\text{V}$$
$$S = \frac{\Delta\text{Output}}{\Delta d}$$
$$= \frac{4}{2}\,\text{V/cm}$$
$$= 2\,\text{V/cm} \tag{3.9}$$

The sensitivity of the proximity sensor is $2\,\text{V/cm}$.

3.4 Applications in Robotics

Proximity sensors find extensive applications in robotics for object detection and collision avoidance. These sensors help robots navigate their environment safely and efficiently.

### 3.4.1	Robot Collision Avoidance

In a robotic system, proximity sensors are often used to detect obstacles in the robot's path. The robot's control system can be programmed to respond by altering the robot's trajectory or slowing down to avoid collisions.

### 3.4.2	Example: Collision Avoidance System

Consider a robot equipped with four proximity sensors on its periphery. If the sensors detect an object within $20\,\text{cm}$ of the robot, the control system triggers a collision avoidance maneuver. The control algorithm can be represented as:

$$
\text{Action} = \begin{cases} \text{Avoidance Maneuver}, & \text{if } d < 20\,\text{cm} \\ \text{Normal Operation}, & \text{otherwise} \end{cases}
\tag{3.10}
$$

This algorithm ensures the robot takes appropriate action based on the proximity sensor readings.

## 3.5	Light Sensors

Light sensors, also known as photodetectors, are devices that convert incident light into an electrical signal. One common type is the photodiode, which operates based on the photoelectric effect. The relationship between the generated photocurrent (I_{ph}), incident light intensity (P_{in}), and the responsivity (R) can be expressed as:

$$
I_{\text{ph}} = R \cdot P_{\text{in}}
\tag{3.11}
$$

where I_{ph} is the photocurrent, P_{in} is the incident light power, and R is the responsivity.

### 3.5.1	Example: Photodiode Responsivity

Consider a silicon photodiode with a responsivity (R) of $0.6\,\text{A/W}$ exposed to an incident light power (P_{in}) of $5\,\text{mW}$. Calculate the resulting photocurrent (I_{ph}).

$$R = 0.6\,\mathrm{A/W}$$

$$P_{\mathrm{in}} = 5\,\mathrm{mW} = 0.005\,\mathrm{W}$$

$$I_{\mathrm{ph}} = R \cdot P_{\mathrm{in}}$$

$$= 0.6\,\mathrm{A/W} \cdot 0.005\,\mathrm{W}$$

$$= 0.003\,\mathrm{A} = 3\,\mathrm{mA} \tag{3.12}$$

The resulting photocurrent is $3\,\mathrm{mA}$.

3.5.2 Solar Cells and Photovoltaic Power

Light sensors are fundamental components in solar cells, converting sunlight into electrical power through the photovoltaic effect. The power generated (P_{out}) by a solar cell can be determined by the product of the incident light power (P_{in}), the area of the solar cell (A), and the efficiency (η):

$$P_{\mathrm{out}} = \eta \cdot P_{\mathrm{in}} \cdot A \tag{3.13}$$

3.5.3 Example: Solar Cell Power

Consider a solar cell with an efficiency (η) of 18% and an incident light power (P_{in}) of $500\,\mathrm{mW}$ over an area (A) of $0.02\,\mathrm{m}^2$. Calculate the generated power (P_{out}).

$$\eta = 0.18$$

$$P_{\mathrm{in}} = 500\,\mathrm{mW} = 0.5\,\mathrm{W}$$

$$A = 0.02\,\mathrm{m}^2$$

$$P_{\mathrm{out}} = \eta \cdot P_{\mathrm{in}} \cdot A$$

$$= 0.18 \cdot 0.5 \cdot 0.02$$

$$= 0.0018\,\mathrm{W} = 1.8\,\mathrm{mW} \tag{3.14}$$

The generated power is $1.8\,\text{mW}$.

3.6   Light Sensors in Ambient Light Detection

Light sensors are widely used in electronic devices for ambient light detection, adjusting display brightness accordingly. The lux level (E) can be related to the illuminance (I) and the area (A) using the formula:

$$E = \frac{I}{A} \tag{3.15}$$

3.6.1   Example: Ambient Light Detection

Consider a light sensor with an illuminance (I) of $200\,\text{lux}$ over an area (A) of $0.01\,\text{m}^2$. Calculate the lux level (E).

$$
\begin{aligned}
I &= 200\,\text{lux} \\
A &= 0.01\,\text{m}^2 \\
E &= \frac{I}{A} \\
&= \frac{200}{0.01} \\
&= 20000\,\text{lux}
\end{aligned}
\tag{3.16}
$$

The lux level is $20000\,\text{lux}$.

Chapter 4

Fiber Optics

4.1 Fiber Optic Basics

Fiber optics is a technology that utilizes thin, flexible strands of glass or plastic known as optical fibers to transmit data using light signals. The basic components of a fiber optic system include the core, cladding, and a protective outer layer. The relationship between the core diameter (D), numerical aperture (NA), and critical angle (θ_c) can be expressed using the numerical aperture formula:

$$NA = \sqrt{n_1^2 - n_2^2} \tag{4.1}$$

where n_1 is the refractive index of the core, and n_2 is the refractive index of the cladding.

4.1.1 Example: Numerical Aperture Calculation

Consider a fiber optic system with a core refractive index (n_1) of 1.5 and a cladding refractive index (n_2) of 1.48. Calculate the numerical aperture (NA).

$$n_1 = 1.5$$

$$n_2 = 1.48$$

$$NA = \sqrt{1.5^2 - 1.48^2}$$

$$\approx 0.36 \tag{4.2}$$

The numerical aperture is approximately 0.36.

4.1.2 Total Internal Reflection

One of the key principles of fiber optics is total internal reflection, which occurs when light traveling through the core is incident at an angle greater than the critical angle (θ_c). The critical angle can be determined using Snell's Law:

$$\theta_c = \sin^{-1}\left(\frac{n_2}{n_1}\right) \tag{4.3}$$

4.1.3 Example: Critical Angle Calculation

For a fiber optic system with a core refractive index (n_1) of 1.5 and a cladding refractive index (n_2) of 1.48, calculate the critical angle (θ_c).

$$n_1 = 1.5$$

$$n_2 = 1.48$$

$$\theta_c = \sin^{-1}\left(\frac{1.48}{1.5}\right)$$

$$\approx 11.54^\circ \tag{4.4}$$

The critical angle is approximately 11.54°.

4.1.4 Propagation of Light in Optical Fibers

In a fiber optic system, light signals propagate through the core via multiple internal reflections, maintaining signal integrity over long distances. The relationship between the number of reflections (N), fiber length (L), and core diameter (D) can be expressed using the formula:

$$N = \frac{L}{\pi D} \tag{4.5}$$

4.1.5 Example: Reflections in Optical Fiber

Consider a single-mode optical fiber with a length (L) of 10 km and a core diameter (D) of 8 μm. Calculate the number of reflections (N).

$$L = 10\,\text{km} = 10^4\,\text{m}$$
$$D = 8\,\mu\text{m} = 8 \times 10^{-6}\,\text{m}$$
$$N = \frac{10^4}{\pi \times 8 \times 10^{-6}}$$
$$\approx 397887\,\text{reflections} \tag{4.6}$$

The number of reflections is approximately 397887.

4.2 Applications of Fiber Optics

Fiber optics find extensive applications in telecommunications, data transmission, and medical imaging. The high bandwidth and low signal loss make fiber optics a preferred choice in various industries.

4.2.1 Telecommunications

In telecommunications, fiber optics form the backbone of high-speed data transmission networks. The ability to transmit data over long distances with minimal signal loss makes fiber optics essential for global communication infrastructure.

4.2.2 Example: Bandwidth Calculation

Consider a single-mode fiber optic system with a bandwidth (B) of 10 Gbps and a wavelength (λ) of 1550 nm. Calculate the data rate capacity using the formula:

$$\text{Data Rate} = B \times \log_2\left(1 + \frac{S}{N}\right) \tag{4.7}$$

where S is the signal power and N is the noise power.

$$B = 10\,\text{Gbps}$$

$$\lambda = 1550\,\text{nm}$$

$$\text{Data Rate} = 10 \times \log_2\left(1 + \frac{S}{N}\right)$$

$$\approx 10 \times \log_2\left(1 + \frac{1\,\text{mW}}{0.1\,\text{mW}}\right)$$

$$\approx 10 \times \log_2(11)$$

$$\approx 34.42\,\text{Gbps} \tag{4.8}$$

The data rate capacity is approximately 34.42 Gbps.

4.2.3 Medical Imaging

Fiber optics play a crucial role in medical imaging, enabling the transmission of light signals for endoscopes and other diagnostic devices. The flexibility and small size of optical fibers make them suitable for minimally invasive medical procedures.

4.2.4 Example: Endoscope Fiber Length

Consider an endoscope with a fiber optic cable of length (L) 2 m. If the core diameter (D) is 150 μm, calculate the number of internal reflections (N) during light propagation.

$$L = 2\,\mathrm{m}$$

$$D = 150\,\mu\mathrm{m} = 150 \times 10^{-6}\,\mathrm{m}$$

$$N = \frac{2}{\pi \times 150 \times 10^{-6}}$$

$$\approx 4242\,\text{reflections} \tag{4.9}$$

The number of internal reflections is approximately 4242.

4.3 Optical Fiber Communication

Optical fiber communication involves the transmission of information using light signals through optical fibers. The performance of an optical fiber communication system is influenced by various parameters, including the optical power (P_{in}), fiber length (L), and attenuation (α). The relationship between these parameters can be expressed using the formula:

$$P_{\mathrm{out}} = P_{\mathrm{in}} \cdot e^{-\alpha L} \tag{4.10}$$

where P_{out} is the output power.

4.3.1 Example: Signal Attenuation

Consider an optical fiber with an input power (P_{in}) of 10 mW, a fiber length (L) of 5 km, and an attenuation coefficient (α) of 0.2 dB/km. Calculate the output power (P_{out}).

$$P_{\text{in}} = 10\,\text{mW}$$

$$L = 5\,\text{km} = 5000\,\text{m}$$

$$\alpha = 0.2\,\text{dB/km}$$

$$P_{\text{out}} = 10 \cdot e^{-0.2 \times 5}$$

$$\approx 10 \cdot e^{-1}$$

$$\approx 3.678\,\text{mW} \tag{4.11}$$

The output power is approximately $3.678\,\text{mW}$.

4.3.2   Chromatic Dispersion in Optical Fibers

Chromatic dispersion is a phenomenon in optical fibers where different wavelengths of light travel at different speeds, causing signal distortion. The dispersion (Δt) can be calculated using the formula:

$$\Delta t = \frac{D \cdot L}{c} \tag{4.12}$$

where D is the dispersion coefficient, L is the fiber length, and c is the speed of light.

4.3.3   Example: Chromatic Dispersion

Consider a single-mode optical fiber with a dispersion coefficient (D) of 17 ps/(nm km), a fiber length (L) of 100 km, and the speed of light (c) as 3×10^8 m/s. Calculate the chromatic dispersion (Δt).

$$D = 17\,\text{ps}/(\text{nm km})$$

$$L = 100\,\text{km} = 100000\,\text{m}$$

$$c = 3 \times 10^8\,\text{m/s}$$

$$\Delta t = \frac{17 \times 100000}{3 \times 10^8}$$

$$\approx 5.67\,\text{ms} \tag{4.13}$$

The chromatic dispersion is approximately $5.67\,\text{ms}$.

4.3.4 Optical Amplification

In optical fiber communication systems, optical amplifiers are used to compensate for signal loss. The gain (G) of an optical amplifier can be determined using the formula:

$$G = 10 \cdot \log_{10}\left(\frac{P_{\text{out}}}{P_{\text{in}}}\right) \tag{4.14}$$

4.3.5 Example: Optical Amplifier Gain

Consider an optical fiber communication system with an output power (P_{out}) of $15\,\text{mW}$ and an input power (P_{in}) of $10\,\text{mW}$. Calculate the gain (G) of the optical amplifier.

$$P_{\text{out}} = 15\,\text{mW}$$

$$P_{\text{in}} = 10\,\text{mW}$$

$$G = 10 \cdot \log_{10}\left(\frac{15}{10}\right)$$

$$\approx 1.76\,\text{dB} \tag{4.15}$$

The gain of the optical amplifier is approximately $1.76\,\text{dB}$.

4.4 Applications of Optical Fiber Communication

Optical fiber communication has revolutionized long-distance communication, enabling high-speed data transmission and reliable connectivity. The applications of optical fiber communication are vast and continue to evolve.

4.4.1 Long-Distance Telecommunication

One of the primary applications of optical fiber communication is in long-distance telecommunication networks. The low signal loss and high bandwidth of optical fibers make them ideal for transmitting voice, data, and video over vast distances.

4.4.2 Example: Transmission Distance

Consider an optical fiber communication link with an attenuation coefficient (α) of $0.3\,\text{dB/km}$ and an input power (P_{in}) of $20\,\text{mW}$. Determine the maximum transmission distance (L_{max}) that ensures the output power (P_{out}) remains above $5\,\text{mW}$.

$$
\alpha = 0.3\,\text{dB/km}
$$

$$
P_{\text{in}} = 20\,\text{mW}
$$

$$
P_{\text{out}} = 5\,\text{mW}
$$

$$
\begin{aligned}
L_{\text{max}} &= \frac{\log_{10}\left(\frac{P_{\text{out}}}{P_{\text{in}}}\right)}{\alpha} \\
&\approx \frac{\log_{10}\left(\frac{5}{20}\right)}{0.3} \\
&\approx 11.55\,\text{km}
\end{aligned}
\tag{4.16}
$$

The maximum transmission distance is approximately $11.55\,\text{km}$.

4.4.3 High-Speed Internet

Optical fiber communication is the backbone of high-speed internet connectivity. Fiber optic cables enable the seamless transmission of large volumes of data, supporting high-speed internet services for businesses and individuals.

4.4.4 Example: Data Rate Capacity

Consider a fiber optic communication system with a bandwidth (B) of 40 Gbps and a wavelength (λ) of 1550 nm. Calculate the data rate capacity using the formula:

$$\text{Data Rate} = B \times \log_2\left(1 + \frac{S}{N}\right) \tag{4.17}$$

where S is the signal power and N is the noise power.

$$B = 40\,\text{Gbps}$$

$$\lambda = 1550\,\text{nm}$$

$$\text{Data Rate} = 40 \times \log_2\left(1 + \frac{1\,\text{mW}}{0.1\,\text{mW}}\right)$$

$$\approx 40 \times \log_2(11)$$

$$\approx 137.5\,\text{Gbps} \tag{4.18}$$

The data rate capacity is approximately 137.5 Gbps.

Chapter 5

Optical Amplifiers

In this chapter, we explore the fascinating world of Optical Amplifiers, with a specific focus on Semiconductor Optical Amplifiers (SOAs). Optical amplifiers play a crucial role in boosting optical signals, enabling efficient long-distance communication in optical fiber networks. Understanding the principles and formulas associated with semiconductor optical amplifiers is essential for professionals in the field of optoelectronics.

5.1 Semiconductor Optical Amplifiers

Semiconductor Optical Amplifiers (SOAs) are devices that use semiconductor materials to amplify optical signals. They find applications in optical communication systems, providing gain to signals in the form of increased optical power. The gain (G) of an SOA can be determined using the formula:

$$G = \frac{\Delta N}{\Delta N_{\text{sat}}} \cdot L \cdot \sigma \tag{5.1}$$

where ΔN is the change in carrier density, ΔN_{sat} is the saturation carrier density, L is the length of the SOA, and σ is the cross-sectional area of the active region.

5.1.1 Example: SOA Gain Calculation

Consider an SOA with a change in carrier density (ΔN) of $2 \times 10^{18}\,\mathrm{cm^{-3}}$, a saturation carrier density (ΔN_{sat}) of $5 \times 10^{18}\,\mathrm{cm^{-3}}$, an SOA length ($L$) of $1\,\mathrm{mm}$, and a cross-sectional area (σ) of $2 \times 10^{-12}\,\mathrm{cm^2}$. Calculate the gain ($G$).

$$\Delta N = 2 \times 10^{18}\,\mathrm{cm^{-3}}$$

$$\Delta N_{\mathrm{sat}} = 5 \times 10^{18}\,\mathrm{cm^{-3}}$$

$$L = 1\,\mathrm{mm} = 0.1\,\mathrm{cm}$$

$$\sigma = 2 \times 10^{-12}\,\mathrm{cm^2}$$

$$G = \frac{2 \times 10^{18}}{5 \times 10^{18}} \cdot 0.1 \cdot 2 \times 10^{-12}$$

$$= 0.2 \times 10^{-12}\,\mathrm{cm} \tag{5.2}$$

The gain of the SOA is $0.2 \times 10^{-12}\,\mathrm{cm}$.

5.1.2 Saturation Intensity

The saturation intensity (I_{sat}) of an SOA represents the intensity at which the gain saturates. It can be calculated using the formula:

$$I_{\mathrm{sat}} = \frac{P_{\mathrm{sat}}}{A} \tag{5.3}$$

where P_{sat} is the saturation power and A is the cross-sectional area of the active region.

5.1.3 Example: Saturation Intensity

For an SOA with a saturation power (P_{sat}) of $1\,\mathrm{mW}$ and a cross-sectional area (A) of $2 \times 10^{-12}\,\mathrm{cm^2}$, calculate the saturation intensity (I_{sat}).

$$P_{\text{sat}} = 1\,\text{mW} = 0.001\,\text{W}$$

$$A = 2 \times 10^{-12}\,\text{cm}^2$$

$$I_{\text{sat}} = \frac{0.001}{2 \times 10^{-12}}$$

$$= 5 \times 10^8\,\text{W/cm}^2 \tag{5.4}$$

The saturation intensity is $5 \times 10^8\,\text{W/cm}^2$.

5.1.4 Gain Saturation

The gain saturation in an SOA occurs when the input signal power (P_{in}) exceeds the saturation power (P_{sat}). The saturated gain (G_{sat}) can be calculated using the formula:

$$G_{\text{sat}} = \frac{G}{1 + \frac{P_{\text{in}}}{P_{\text{sat}}}} \tag{5.5}$$

5.1.5 Example: Gain Saturation

For an SOA with a gain (G) of $0.2 \times 10^{-12}\,\text{cm}$, a saturation power ($P_{\text{sat}}$) of $1\,\text{mW}$, and an input power (P_{in}) of $2\,\text{mW}$, calculate the saturated gain (G_{sat}).

$$G = 0.2 \times 10^{-12}\,\text{cm}$$

$$P_{\text{sat}} = 1\,\text{mW}$$

$$P_{\text{in}} = 2\,\text{mW}$$

$$G_{\text{sat}} = \frac{0.2 \times 10^{-12}}{1 + \frac{2}{1}}$$

$$= 0.1 \times 10^{-12}\,\text{cm} \tag{5.6}$$

The saturated gain is $0.1 \times 10^{-12}\,\text{cm}$.

5.2   Applications of Semiconductor Optical Amplifiers

Semiconductor Optical Amplifiers (SOAs) are integral components in optical communication systems, finding applications in various domains.

5.2.1   Wavelength Conversion

SOAs are used for wavelength conversion in optical communication systems, allowing the conversion of signals from one wavelength to another.

5.2.2   Example: Wavelength Conversion Efficiency

Consider an SOA used for wavelength conversion with an input power (P_{in}) of $5\,\text{mW}$ and an output power (P_{out}) of $10\,\text{mW}$. Calculate the wavelength conversion efficiency (η) using the formula:

$$\eta = \frac{P_{\text{out}} - P_{\text{in}}}{P_{\text{in}}} \tag{5.7}$$

$$
\begin{aligned}
P_{\text{in}} &= 5\,\text{mW} \\
P_{\text{out}} &= 10\,\text{mW} \\
\eta &= \frac{10 - 5}{5} \\
&= 1 \tag{5.8}
\end{aligned}
$$

The wavelength conversion efficiency is 1 or 100%.

5.2.3   Signal Amplification

SOAs are primarily used for signal amplification in optical communication systems, enhancing the strength of optical signals.

5.2.4 Example: Amplification Factor

For an SOA with an input power (P_{in}) of $3\,\text{mW}$ and an output power (P_{out}) of $15\,\text{mW}$, calculate the amplification factor (AF) using the formula:

$$\text{AF} = \frac{P_{\text{out}}}{P_{\text{in}}} \tag{5.9}$$

$$P_{\text{in}} = 3\,\text{mW}$$

$$P_{\text{out}} = 15\,\text{mW}$$

$$\text{AF} = \frac{15}{3}$$

$$= 5 \tag{5.10}$$

The amplification factor is 5.

5.3 Erbium-Doped Fiber Amplifiers

Erbium-Doped Fiber Amplifiers (EDFAs) are optical amplifiers that utilize erbium-doped fibers to amplify optical signals. The gain (G) of an EDFA can be calculated using the formula:

$$G = \frac{\sigma \cdot N \cdot L}{A_{\text{eff}}} \tag{5.11}$$

where σ is the emission cross-section of erbium, N is the population inversion, L is the length of the erbium-doped fiber, and A_{eff} is the effective core area.

5.3.1 Example: EDFA Gain Calculation

Consider an EDFA with an emission cross-section (σ) of $4 \times 10^{-25}\,\text{cm}^2$, a population inversion (N) of $1.5 \times 10^{25}\,\text{cm}^{-3}$, an erbium-doped fiber length (L) of $10\,\text{m}$, and an effective core area (A_{eff}) of $80\,\eta\text{m}^2$. Calculate the gain (G).

$$\sigma = 4 \times 10^{-25} \, \text{cm}^2$$

$$N = 1.5 \times 10^{25} \, \text{cm}^{-3}$$

$$L = 10 \, \text{m}$$

$$A_{\text{eff}} = 80 \, \eta \, \text{m}^2 = 8 \times 10^{-8} \, \text{cm}^2$$

$$G = \frac{4 \times 10^{-25} \cdot 1.5 \times 10^{25} \cdot 10}{8 \times 10^{-8}}$$

$$= 7.5 \times 10^6 \tag{5.12}$$

The gain of the EDFA is 7.5×10^6.

5.3.2 Saturation Power

The saturation power (P_{sat}) of an EDFA is the input power level at which the gain saturates. It can be determined using the formula:

$$P_{\text{sat}} = \frac{h \cdot \nu \cdot A_{\text{eff}} \cdot \gamma}{\sigma \cdot \tau_{\text{rad}}} \tag{5.13}$$

where h is Planck's constant, ν is the frequency, γ is the confinement factor, τ_{rad} is the radiative lifetime.

5.3.3 Example: Saturation Power

For an EDFA with Planck's constant (h) of 6.626×10^{-34} Joule s, a frequency (ν) of 193×10^{12} Hz, a confinement factor (γ) of 0.8, and a radiative lifetime (τ_{rad}) of 10 ms, calculate the saturation power (P_{sat}).

$$h = 6.626 \times 10^{-34} \text{ Joule s}$$

$$\nu = 193 \times 10^{12} \text{ Hz}$$

$$\gamma = 0.8$$

$$\tau_{\text{rad}} = 10\,\text{ms} = 0.01\,\text{s}$$

$$\sigma = 4 \times 10^{-25}\,\text{cm}^2 \text{ (from previous example)}$$

$$A_{\text{eff}} = 8 \times 10^{-8}\,\text{cm}^2 \text{ (from previous example)}$$

$$P_{\text{sat}} = \frac{6.626 \times 10^{-34} \cdot 193 \times 10^{12} \cdot 8 \times 10^{-8} \cdot 0.8}{4 \times 10^{-25} \cdot 0.01}$$

$$\approx 2.05 \times 10^{-9}\,\text{W} \tag{5.14}$$

The saturation power is approximately 2.05×10^{-9} W.

5.3.4 Gain Saturation

The gain saturation in an EDFA occurs when the input power (P_{in}) exceeds the saturation power (P_{sat}). The saturated gain (G_{sat}) can be calculated using the formula:

$$G_{\text{sat}} = \frac{G}{1 + \dfrac{P_{\text{in}}}{P_{\text{sat}}}} \tag{5.15}$$

5.3.5 Example: Gain Saturation

For an EDFA with a gain (G) of 7.5×10^6, a saturation power (P_{sat}) of 2.05×10^{-9} W, and an input power (P_{in}) of 5×10^{-9} W, calculate the saturated gain (G_{sat}).

$$G = 7.5 \times 10^6 \text{ (from previous example)}$$

$$P_{\text{sat}} = 2.05 \times 10^{-9}\,\text{W (from previous example)}$$

$$P_{\text{in}} = 5 \times 10^{-9}\,\text{W}$$

$$G_{\text{sat}} = \frac{7.5 \times 10^6}{1 + \frac{5 \times 10^{-9}}{2.05 \times 10^{-9}}}$$

$$\approx 3.06 \times 10^6 \tag{5.16}$$

The saturated gain is approximately 3.06×10^6.

5.4 Applications of Erbium-Doped Fiber Amplifiers

Erbium-Doped Fiber Amplifiers (EDFAs) are widely used in optical communication systems for various applications.

5.4.1 Long-Haul Optical Communication

EDFAs play a crucial role in long-haul optical communication systems, where signals need to be amplified over extended distances.

5.4.2 Example: Signal Power at Receiver

Consider an optical communication system using EDFAs with an input power (P_{in}) of $10\,\eta\text{W}$ and a gain (G) of 7.5×10^6. Calculate the signal power at the receiver (P_{out}).

$$P_{\text{out}} = P_{\text{in}} \times G \tag{5.17}$$

$$P_{\text{in}} = 10\,\eta\text{W} = 10^{-5}\,\text{W}$$

$$G = 7.5 \times 10^6 \text{ (from previous example)}$$

$$P_{\text{out}} = 10^{-5} \times 7.5 \times 10^6$$

$$= 75\,\text{W} \tag{5.18}$$

The signal power at the receiver is 75 W.

5.4.3 Wavelength Division Multiplexing (WDM)

EDFAs are essential components in Wavelength Division Multiplexing (WDM) systems, where multiple signals are transmitted simultaneously at different wavelengths.

5.4.4 Example: WDM Channel Separation

Consider a WDM system using EDFAs with five channels at wavelengths (λ) of 1550 nm, 1560 nm, 1570 nm, 1580 nm, and 1590 nm. If the channel separation is 20 nm, calculate the channel separation in frequency (Δf).

$$\Delta f = \frac{c}{\lambda} \tag{5.19}$$

where c is the speed of light.

$$c = 3 \times 10^8\,\text{m/s}$$

$$\lambda = 20 \times 10^{-9}\,\text{m}$$

$$\Delta f = \frac{3 \times 10^8}{20 \times 10^{-9}}$$

$$= 1.5 \times 10^{13}\,\text{Hz} \tag{5.20}$$

The channel separation in frequency is 1.5×10^{13} Hz.

Chapter 6

Optoelectronic Devices in Communication

6.1 Lasers

Lasers, or Light Amplification by Stimulated Emission of Radiation, are devices that produce coherent and focused light through a process of stimulated emission. The fundamental equation governing the energy levels in a laser medium is given by:

$$E = \frac{hc}{\lambda} \tag{6.1}$$

where E is the energy, h is Planck's constant, c is the speed of light, and λ is the wavelength.

6.1.1 Example: Calculating Photon Energy

Consider a laser emitting light at a wavelength (λ) of 800 nm. Calculate the energy of a single photon using Planck's constant ($h = 6.626 \times 10^{-34}$ Joule s).

$$h = 6.626 \times 10^{-34} \text{ Joule s}$$

$$c = 3 \times 10^8 \text{ m/s}$$

$$\lambda = 800 \times 10^{-9} \text{ m}$$

$$E = \frac{6.626 \times 10^{-34} \times 3 \times 10^8}{800 \times 10^{-9}}$$

$$\approx 2.48 \times 10^{-16} \text{ Joule} \tag{6.2}$$

The energy of a single photon is approximately 2.48×10^{-16} Joule.

6.1.2 Population Inversion

Achieving population inversion, where more atoms are in higher energy states than lower ones, is crucial for laser operation. The population inversion factor (P) is given by:

$$P = \frac{N_2 - N_1}{N} \tag{6.3}$$

where N_2 is the number of atoms in the higher energy state, N_1 is the number of atoms in the lower energy state, and N is the total number of atoms.

6.1.3 Example: Population Inversion Factor

Consider a laser medium with $N_2 = 8 \times 10^{23}$, $N_1 = 2 \times 10^{23}$, and $N = 1 \times 10^{24}$. Calculate the population inversion factor (P).

$$N_2 = 8 \times 10^{23}$$

$$N_1 = 2 \times 10^{23}$$

$$N = 1 \times 10^{24}$$

$$P = \frac{8 \times 10^{23} - 2 \times 10^{23}}{1 \times 10^{24}}$$

$$= 0.6 \tag{6.4}$$

The population inversion factor is 0.6.

6.1.4 Laser Gain

The gain (G) in a laser medium is proportional to the product of the population inversion factor and the stimulated emission cross-section (σ):

$$G = \sigma \cdot P \tag{6.5}$$

6.1.5 Example: Laser Gain

For a laser medium with a population inversion factor (P) of 0.6 and a stimulated emission cross-section (σ) of $2 \times 10^{-19}\,\mathrm{cm}^2$, calculate the laser gain (G).

$$P = 0.6$$
$$\sigma = 2 \times 10^{-19}\,\mathrm{cm}^2$$
$$G = 2 \times 10^{-19} \times 0.6$$
$$= 1.2 \times 10^{-19}\,\mathrm{cm}^2 \tag{6.6}$$

The laser gain is $1.2 \times 10^{-19}\,\mathrm{cm}^2$.

6.1.6 Threshold Gain

The threshold gain (G_{th}) is the gain required to achieve laser oscillation. It is inversely proportional to the cavity length (L):

$$G_{\mathrm{th}} = \frac{\pi}{L} \tag{6.7}$$

6.1.7 Example: Threshold Gain

For a laser with a cavity length (L) of 0.1 m, calculate the threshold gain (G_{th}).

$$L = 0.1\,\text{m}$$

$$G_{\text{th}} = \frac{\pi}{0.1}$$

$$= 31.83\,\text{cm}^{-1} \tag{6.8}$$

The threshold gain is $31.83\,\text{cm}^{-1}$.

6.2   Applications of Lasers

Lasers find extensive applications in various fields, and their unique properties make them indispensable in optical communication.

6.2.1   Fiber Optic Communication

Lasers are widely used as light sources in fiber optic communication systems, ensuring high-speed and reliable data transmission.

6.2.2   Example: Data Rate Calculation

Consider a fiber optic communication system using a laser with a wavelength (λ) of 1550 nm and a data rate capacity (B) of 40 Gbps. Calculate the data rate capacity using the formula:

$$\text{Data Rate} = B \times \log_2\left(1 + \frac{S}{N}\right) \tag{6.9}$$

where S is the signal power and N is the noise power.

$$B = 40\,\text{Gbps}$$

$$\lambda = 1550\,\text{nm}$$

$$\text{Data Rate} = 40 \times \log_2\left(1 + \frac{1\,\text{mW}}{0.1\,\text{mW}}\right)$$

$$\approx 40 \times \log_2(11)$$

$$\approx 137.5\,\text{Gbps} \tag{6.10}$$

The data rate capacity is approximately 137.5 Gbps.

6.2.3 Medical Applications

Lasers are extensively used in medical applications, such as laser surgery and diagnostics, owing to their precision and minimal invasiveness.

6.2.4 Example: Tissue Ablation

In laser surgery, the depth of tissue ablation (D) can be calculated using the formula:

$$D = \frac{\lambda}{4n} \tag{6.11}$$

where λ is the laser wavelength and n is the refractive index of the tissue.

$$\lambda = 1064\,\text{nm}$$

$$n = 1.4$$

$$D = \frac{1064}{4 \times 1.4}$$

$$\approx 190\,\text{nm} \tag{6.12}$$

The depth of tissue ablation is approximately 190 nm.

6.3   Optical Modulators

Optical modulators are devices that control the intensity, phase, or polarization of light signals, allowing for the encoding of information onto optical carriers. The modulation index (m) of an optical modulator, which represents the extent of modulation, can be calculated using the formula:

$$m = \frac{A_{\text{max}} - A_{\text{min}}}{A_{\text{max}} + A_{\text{min}}} \tag{6.13}$$

where A_{max} is the maximum amplitude, and A_{min} is the minimum amplitude.

6.3.1   Example: Modulation Index Calculation

Consider an optical modulator with a maximum amplitude (A_{max}) of 2 V and a minimum amplitude (A_{min}) of 0.5 V. Calculate the modulation index (m).

$$A_{\text{max}} = 2\,\text{V}$$

$$A_{\text{min}} = 0.5\,\text{V}$$

$$m = \frac{2 - 0.5}{2 + 0.5}$$

$$= 0.4286 \tag{6.14}$$

The modulation index is approximately 0.4286.

6.3.2   Frequency Modulation (FM)

Frequency modulation is a common modulation technique where the frequency of the optical carrier signal is varied in accordance with the message signal. The frequency deviation (Δf) in an FM modulator can be determined using the formula:

$$\Delta f = f_{\text{max}} - f_{\text{min}} \tag{6.15}$$

where f_{max} is the maximum frequency and f_{min} is the minimum frequency.

6.3.3 Example: Frequency Deviation

For an FM modulator with a maximum frequency (f_{max}) of 2 GHz and a minimum frequency (f_{min}) of 1.5 GHz, calculate the frequency deviation (Δf).

$$f_{\text{max}} = 2\,\text{GHz}$$

$$f_{\text{min}} = 1.5\,\text{GHz}$$

$$\Delta f = 2 - 1.5$$

$$= 0.5\,\text{GHz} \tag{6.16}$$

The frequency deviation is 0.5 GHz.

6.3.4 Phase Modulation (PM)

Phase modulation involves varying the phase of the optical carrier signal in response to the message signal. The phase deviation ($\Delta\phi$) in a PM modulator is given by:

$$\Delta\phi = \phi_{\text{max}} - \phi_{\text{min}} \tag{6.17}$$

where ϕ_{max} is the maximum phase and ϕ_{min} is the minimum phase.

6.3.5 Example: Phase Deviation

For a PM modulator with a maximum phase (ϕ_{max}) of $45°$ and a minimum phase (ϕ_{min}) of $30°$, calculate the phase deviation ($\Delta\phi$).

$$\phi_{\text{max}} = 45°$$

$$\phi_{\text{min}} = 30°$$

$$\Delta\phi = 45 - 30$$

$$= 15° \tag{6.18}$$

The phase deviation is $15°$.

6.3.6 Electro-Optic Modulators

Electro-optic modulators use the electro-optic effect to modulate the refractive index of a material, influencing the phase of the transmitted light. The modulation depth (Δn) in an electro-optic modulator is given by:

$$\Delta n = r \cdot E \tag{6.19}$$

where r is the electro-optic coefficient and E is the applied electric field.

6.3.7 Example: Modulation Depth

For an electro-optic modulator with an electro-optic coefficient (r) of 2×10^{-12} m/V and an applied electric field (E) of $100\,$kV/m, calculate the modulation depth (Δn).

$$r = 2 \times 10^{-12}\,\text{m/V}$$

$$E = 100 \times 10^3\,\text{V/m}$$

$$\Delta n = 2 \times 10^{-12} \times 100 \times 10^3$$

$$= 2 \times 10^{-7}\,\text{m} \tag{6.20}$$

The modulation depth is 2×10^{-7} m.

6.4 Applications of Optical Modulators

Optical modulators find widespread applications in optical communication systems, facilitating the transmission of information through various modulation techniques.

6.4.1 Quadrature Amplitude Modulation (QAM)

QAM is a modulation scheme that combines both amplitude and phase modulation. The modulation constellations (M) in a QAM system can be determined using the formula:

$$M = 2^{\frac{\text{Bits}}{2}} \tag{6.21}$$

6.4.2 Example: QAM Modulation Constellations

For a QAM system with 16 bits, calculate the modulation constellations (M).

$$M = 2^{\frac{16}{2}} = 2^8 = 256 \tag{6.22}$$

The QAM modulation constellations are 256.

6.4.3 Optical Frequency Division Multiplexing (OFDM)

OFDM is a multiplexing technique that divides the total bandwidth into multiple orthogonal subcarriers. The subcarrier spacing (Δf) in an OFDM system is given by:

$$\Delta f = \frac{1}{T} \tag{6.23}$$

where T is the symbol period.

6.4.4 Example: Subcarrier Spacing

For an OFDM system with a symbol period (T) of $1\,\eta$s, calculate the subcarrier spacing (Δf).

$$T = 1 \times 10^{-6}\,\text{s}$$
$$\Delta f = \frac{1}{1 \times 10^{-6}}$$
$$= 1\,\text{MHz} \tag{6.24}$$

The subcarrier spacing is $1\,\text{MHz}$.

Chapter 7

Optoelectronic Integrated Circuits

7.1 Photonic Integrated Circuits

Photonic Integrated Circuits (PICs) are devices that integrate various optical components, such as lasers, modulators, and detectors, onto a single chip. The effective refractive index (n_{eff}) in a waveguide of a PIC can be calculated using the formula:

$$n_{\text{eff}} = \sqrt{n_{\text{core}}^2 - \frac{2\pi^2}{\lambda^2} \cdot (\text{core width})^2} \tag{7.1}$$

where n_{core} is the refractive index of the core, λ is the wavelength, and the core width is the width of the waveguide core.

7.1.1 Example: Effective Refractive Index

Consider a PIC with a core refractive index (n_{core}) of 3.5, a wavelength (λ) of 1550 nm, and a core width of 2 ηm. Calculate the effective refractive index (n_{eff}).

$$n_{\text{core}} = 3.5$$

$$\lambda = 1550\,\text{nm}$$

$$\text{Core width} = 2\,\mu\text{m}$$

$$n_{\text{eff}} = \sqrt{3.5^2 - \frac{2\pi^2}{(1550 \times 10^{-9})^2} \cdot (2 \times 10^{-6})^2}$$

$$\approx 3.479 \tag{7.2}$$

The effective refractive index is approximately 3.479.

7.1.2　Propagation Delay

The propagation delay (t_{pd}) in a waveguide of a PIC is given by:

$$t_{\text{pd}} = \frac{L}{v_{\text{g}}} \tag{7.3}$$

where L is the length of the waveguide, and v_{g} is the group velocity.

7.1.3　Example: Propagation Delay

For a PIC with a waveguide length (L) of 5 mm and a group velocity (v_{g}) of 2×10^8 m/s, calculate the propagation delay (t_{pd}).

$$L = 5\,\text{mm} = 5 \times 10^{-3}\,\text{m}$$

$$v_{\text{g}} = 2 \times 10^8\,\text{m/s}$$

$$t_{\text{pd}} = \frac{5 \times 10^{-3}}{2 \times 10^8}$$

$$= 2.5 \times 10^{-11}\,\text{s} \tag{7.4}$$

The propagation delay is 2.5×10^{-11} s.

7.1.4 Power Splitter Loss

The power splitter loss (P_{splitter}) in a PIC is given by:

$$P_{\text{splitter}} = 10 \cdot \log_{10}\left(\frac{1}{\text{Number of Ports}}\right) \tag{7.5}$$

7.1.5 Example: Power Splitter Loss

For a power splitter with 4 ports, calculate the power splitter loss (P_{splitter}).

$$\text{Number of Ports} = 4$$
$$P_{\text{splitter}} = 10 \cdot \log_{10}\left(\frac{1}{4}\right)$$
$$= -6\,\text{dB} \tag{7.6}$$

The power splitter loss is $-6\,\text{dB}$.

7.2   Applications of Photonic Integrated Circuits

Photonic Integrated Circuits find applications in various optical communication and sensing systems, providing compact and efficient solutions.

7.2.1   Wavelength Division Multiplexing (WDM)

WDM is a technology that uses multiple wavelengths to transmit data simultaneously over a single optical fiber. PICs play a crucial role in WDM systems, facilitating the integration of multiple lasers and modulators.

7.2.2   Example: Channel Spacing

In a WDM system using a PIC, if the central wavelengths of two adjacent channels are 1550 nm and 1560 nm, calculate the channel spacing ($\Delta\lambda$).

$$\Delta\lambda = \lambda_2 - \lambda_1 \tag{7.7}$$

$$\lambda_2 = 1560\,\text{nm}$$

$$\lambda_1 = 1550\,\text{nm}$$

$$\Delta\lambda = 1560 - 1550$$

$$= 10\,\text{nm} \tag{7.8}$$

The channel spacing is 10 nm.

7.2.3   Optical Gyroscopes

Optical gyroscopes based on PICs are used for precise measurements of rotation rates. The sensitivity (S) of an optical gyroscope is given by:

$$S = \frac{\lambda}{2 \cdot L \cdot n_{\text{eff}}} \tag{7.9}$$

7.2.4   Example: Gyroscope Sensitivity

For an optical gyroscope with a wavelength (λ) of 1550 nm, a waveguide length (L) of 2 cm, and an effective refractive index (n_{eff}) of 3.5, calculate the gyroscope sensitivity (S).

$$\lambda = 1550\,\text{nm}$$

$$L = 2\,\text{cm} = 0.02\,\text{m}$$

$$n_{\text{eff}} = 3.5$$

$$S = \frac{1550}{2 \cdot 0.02 \cdot 3.5}$$

$$\approx 22.14\,\text{Hz/(deg/s)} \tag{7.10}$$

The gyroscope sensitivity is approximately 22.14 Hz/(deg/s).

Photonic Integrated Circuits represent a significant advancement in the field of optoelectronics, enabling the integration of various optical components onto a single chip. The formulas and examples provided in this section offer insights into the principles and applications of Photonic Integrated Circuits, showcasing their importance in optical communication and sensing systems.

7.3 Optical Interconnects

Optical interconnects involve the use of optical signals to transmit data between electronic components, replacing traditional copper-based interconnects. The total link power (P_{total}) in an optical interconnect system can be calculated using the formula:

$$P_{\text{total}} = P_{\text{transmit}} + P_{\text{receive}} \tag{7.11}$$

where P_{transmit} is the power consumed by the transmitter, and P_{receive} is the power consumed by the receiver.

7.3.1 Example: Total Link Power

Consider an optical interconnect system with a transmitter power (P_{transmit}) of 100 mW and a receiver power (P_{receive}) of 80 mW. Calculate the total link power (P_{total}).

$$P_{\text{transmit}} = 100\,\text{mW}$$
$$P_{\text{receive}} = 80\,\text{mW}$$
$$P_{\text{total}} = 100 + 80$$
$$= 180\,\text{mW} \tag{7.12}$$

The total link power is 180 mW.

7.3.2 Bit Error Rate (BER)

The bit error rate (BER) in an optical communication system can be estimated using the formula:

$$\text{BER} = Q\left(\sqrt{\frac{2 \cdot P_{\text{optical}} \cdot \text{BER}_{\text{target}}}{\text{Responsivity} \cdot \text{Bit Rate}}}\right) \tag{7.13}$$

where Q is the Q-function, P_{optical} is the optical power, $\text{BER}_{\text{target}}$ is the target bit error rate, Responsivity is the responsivity of the photodetector, and Bit Rate is the data rate.

7.3.3   Example: Bit Error Rate

For an optical communication system with an optical power (P_{optical}) of -10 dBm, a target bit error rate ($\text{BER}_{\text{target}}$) of 10^{-12}, a responsivity of 0.8 A/W, and a bit rate of 10 Gbps, calculate the bit error rate (BER).

$$P_{\text{optical}} = 10^{-1}\,\text{mW}$$

$$\text{BER}_{\text{target}} = 10^{-12}$$

$$\text{Responsivity} = 0.8\,\text{A/W}$$

$$\text{Bit Rate} = 10\,\text{Gbps}$$

$$\text{BER} = Q\left(\sqrt{\frac{2 \cdot 10^{-1} \cdot 10^{-12}}{0.8 \cdot 10^{10}}}\right)$$

$$\approx Q(4.472 \times 10^{-15}) \tag{7.14}$$

The bit error rate is approximately $Q(4.472 \times 10^{-15})$.

7.3.4   Link Budget

The link budget (LB) in an optical interconnect system is given by:

$$\text{LB} = P_{\text{transmit}} - P_{\text{optical}} - \text{Loss}_{\text{fiber}} - \text{Loss}_{\text{connector}} - \text{Loss}_{\text{splicing}} + \text{Margin} \tag{7.15}$$

where P_{optical} is the optical power, $\text{Loss}_{\text{fiber}}$ is the fiber loss, $\text{Loss}_{\text{connector}}$ is the connector loss, $\text{Loss}_{\text{splicing}}$ is the splicing loss, and Margin is the system margin.

7.3.5 Example: Link Budget

For an optical interconnect system with a transmitter power (P_{transmit}) of 150 mW, an optical power (P_{optical}) of -8 dBm, fiber loss ($\text{Loss}_{\text{fiber}}$) of 2 dB, connector loss ($\text{Loss}_{\text{connector}}$) of 1 dB, splicing loss ($\text{Loss}_{\text{splicing}}$) of 0.5 dB, and a system margin of 3 dB, calculate the link budget (LB).

$$P_{\text{transmit}} = 150\,\text{mW}$$

$$P_{\text{optical}} = 10^{-0.8}\,\text{mW}$$

$$\text{Loss}_{\text{fiber}} = 2\,\text{dB}$$

$$\text{Loss}_{\text{connector}} = 1\,\text{dB}$$

$$\text{Loss}_{\text{splicing}} = 0.5\,\text{dB}$$

$$\text{Margin} = 3\,\text{dB}$$

$$\text{LB} = 150 - 10^{-0.8} - 2 - 1 - 0.5 + 3$$

$$\approx 141.2\,\text{dB} \tag{7.16}$$

The link budget is approximately 141.2 dB.

7.4 Applications of Optical Interconnects

Optical interconnects find applications in various domains, enhancing the performance and efficiency of communication between electronic components.

7.4.1 Data Centers

In data centers, optical interconnects provide high-speed and energy-efficient communication between servers and networking equipment.

7.4.2 Example: Data Center Link

For an optical interconnect in a data center with a link budget (LB) of 140 dB and a receiver sensitivity (Sensitivity) of -25 dBm, calculate the maximum allowable fiber loss.

$$\text{Maximum Allowable Fiber Loss} = \text{LB} - \text{Sensitivity} \tag{7.17}$$

$$\text{LB} = 140\,\text{dB}$$

$$\text{Sensitivity} = -25\,\text{dBm}$$

$$\text{Maximum Allowable Fiber Loss} = 140 - (-25)$$

$$= 165\,\text{dB} \tag{7.18}$$

The maximum allowable fiber loss is $165\,\text{dB}$.

7.4.3   High-Performance Computing (HPC)

In high-performance computing environments, optical interconnects facilitate rapid data transfer between processors and memory units.

7.4.4   Example: HPC System

For an HPC system with a link budget (LB) of 150 dB and a total link power (P_{total}) of 200 mW, calculate the optical power (P_{optical}).

$$P_{\text{optical}} = P_{\text{total}} - \text{LB} \tag{7.19}$$

$$\text{LB} = 150\,\text{dB}$$

$$P_{\text{total}} = 200\,\text{mW}$$

$$P_{\text{optical}} = 200 - 150$$

$$= 50\,\text{mW} \tag{7.20}$$

The optical power is $50\,\text{mW}$.

Chapter 8

Photonic Crystals

In this chapter, we embark on a journey into the fascinating realm of Photonic Crystals. Photonic crystals are periodic structures that exhibit unique optical properties, influencing the propagation of light waves. Understanding the fundamentals of photonic crystals is essential for researchers and engineers working in the field of optoelectronics.

8.1 Introduction to Photonic Crystals

Photonic crystals are artificial structures that manipulate the flow of light in a manner analogous to how semiconductors influence the flow of electrons. The periodic arrangement of dielectric materials in photonic crystals creates bandgaps for specific wavelengths, leading to the inhibition of light propagation in certain directions.

8.1.1 Photonic Crystal Structure

The structure of a one-dimensional photonic crystal can be represented by a sequence of alternating layers with different refractive indices. The characteristic equation for the bandgap in a one-dimensional photonic crystal is given by:

$$\cos\left(\frac{2\pi d}{\lambda}\sqrt{n_1^2 - n_2^2}\right) = \cos\left(\frac{2\pi a}{\lambda}\right) \tag{8.1}$$

where d is the thickness of each layer, a is the periodicity of the crystal, λ is the wavelength, n_1 is the refractive index of the high-index layer, and n_2 is the refractive index of the low-index layer.

8.1.2   Example: Bandgap Calculation

For a one-dimensional photonic crystal with layer thickness (d) of 100 nm, periodicity (a) of 500 nm, refractive indices (n_1) of 3.5 and (n_2) of 1.5, calculate the bandgap wavelength (λ).

$$d = 100\,\text{nm}$$

$$a = 500\,\text{nm}$$

$$n_1 = 3.5$$

$$n_2 = 1.5$$

$$\lambda = \frac{2d}{\sqrt{n_1^2 - n_2^2}}$$

$$= \frac{2 \times 100 \times 10^{-9}}{\sqrt{3.5^2 - 1.5^2}}$$

$$\approx 315.85\,\text{nm} \tag{8.2}$$

The bandgap wavelength is approximately 315.85 nm.

8.1.3   Two-Dimensional Photonic Crystals

For a two-dimensional photonic crystal, the structure consists of a periodic arrangement of high-index and low-index materials in two dimensions. The bandgap conditions are more complex, involving both in-plane and out-of-plane propagation.

8.1.4 Example: Bandgap in Two-Dimensional Photonic Crystal

Consider a two-dimensional photonic crystal with a square lattice and a high-index material (n_1) of 3.8 and a low-index material (n_2) of 1.6. Calculate the bandgap for in-plane propagation with a lattice constant (a) of 400 nm.

$$\cos\left(\frac{2\pi a}{\lambda}\sqrt{n_1^2 - n_2^2}\right) = \cos\left(\frac{\pi a}{\lambda}\right) \tag{8.3}$$

$$
\begin{aligned}
a &= 400\,\text{nm} \\
n_1 &= 3.8 \\
n_2 &= 1.6 \\
\lambda &= \frac{a}{\sqrt{n_1^2 - n_2^2}} \\
&= \frac{400 \times 10^{-9}}{\sqrt{3.8^2 - 1.6^2}} \\
&\approx 462.74\,\text{nm}
\end{aligned}
\tag{8.4}
$$

The bandgap for in-plane propagation is approximately 462.74 nm.

8.2 Applications of Photonic Crystals

Photonic crystals find applications in a myriad of fields, ranging from optical communication to sensing and imaging.

8.2.1 Photonic Crystal Fibers

Photonic crystal fibers utilize the unique properties of photonic crystals to guide light in unconventional ways, enabling improved light confinement and transmission.

8.2.2 Example: Dispersion Engineering

In a photonic crystal fiber designed for dispersion engineering, if the air hole diameter (d) is $2\,\mu$m and the pitch ($\wedge$) is $4\,\mu$m, calculate the normalized frequency (V).

$$V = \frac{2\pi a}{\lambda}\sqrt{n_1^2 - n_2^2} \tag{8.5}$$

$$a = \frac{d}{2}$$

$$\lambda = \frac{\Lambda}{\sqrt{n_1^2 - n_2^2}}$$

$$= \frac{4 \times 10^{-6}}{\sqrt{3.8^2 - 1.6^2}}$$

$$\approx 1500\,\text{nm}$$

$$V = \frac{2\pi \times 1 \times 10^{-6}}{1500 \times 10^{-9}}\sqrt{3.8^2 - 1.6^2}$$

$$\approx 7.46 \tag{8.6}$$

The normalized frequency (V) is approximately 7.46.

8.2.3 Photonic Crystal Sensors

Photonic crystal sensors exploit changes in the band structure of photonic crystals to detect variations in the surrounding environment, offering highly sensitive sensing capabilities.

8.2.4 Example: Sensitivity Calculation

For a photonic crystal sensor with a bandgap wavelength (λ) of $800\,$nm and a refractive index sensitivity (S) of $300\,$nm/RIU, calculate the sensitivity in terms of wavelength shift per refractive index unit.

$$\text{Sensitivity} = \frac{\lambda}{S} \tag{8.7}$$

$$\lambda = 800\,\text{nm}$$

$$S = 300\,\text{nm/RIU}$$

$$\text{Sensitivity} = \frac{800}{300}$$

$$\approx 2.67\,\text{nm/RIU} \tag{8.8}$$

The sensitivity is approximately $2.67\,\text{nm/RIU}$.

8.3 Applications of Photonic Crystals

8.3.1 Optical Communication

Photonic crystals play a pivotal role in the field of optical communication. They are employed to design compact and efficient components such as waveguides and filters, contributing to the miniaturization and improved performance of optical communication systems. The bandgap properties of photonic crystals allow precise control over the propagation of specific wavelengths, enabling the development of wavelength-selective devices.

8.3.2 Example: Photonic Crystal Waveguide

Consider a photonic crystal waveguide designed for a bandgap wavelength (λ) of 1550 nm and a refractive index contrast (Δn) of 0.5. The effective refractive index (n_{eff}) of the waveguide can be calculated using the formula:

$$n_{\text{eff}} = \frac{\lambda}{2a\sqrt{\Delta n}} \tag{8.9}$$

where a is the lattice constant.

$$\lambda = 1550 \, \text{nm}$$

$$\Delta n = 0.5$$

$$a = \frac{\lambda}{2 n_{\text{eff}} \sqrt{\Delta n}}$$

$$= \frac{1550 \times 10^{-9}}{2 \times 1.5 \times \sqrt{0.5}}$$

$$\approx 426 \, \text{nm} \tag{8.10}$$

The lattice constant (a) is approximately 426 nm.

8.3.3   Sensing and Detection

Photonic crystals find applications in highly sensitive sensors and detectors. The ability of photonic crystals to exhibit changes in their optical properties in response to variations in the surrounding environment makes them ideal for sensing applications. The detection of minute changes in refractive index or the presence of specific molecules is achievable through the analysis of bandgap shifts or resonance conditions in photonic crystal structures.

8.3.4   Example: Photonic Crystal Sensor

For a photonic crystal sensor with a lattice constant (a) of 500 nm, a bandgap wavelength (λ) of 800 nm, and a refractive index sensitivity (S) of 400 nm/RIU, calculate the change in refractive index (Δn) required to induce a bandgap shift.

$$\Delta n = \frac{\Delta \lambda}{S} \tag{8.11}$$

$$a = 500\,\text{nm}$$

$$\lambda = 800\,\text{nm}$$

$$S = 400\,\text{nm/RIU}$$

$$\Delta n = \frac{\lambda - a}{S}$$

$$= \frac{800 - 500}{400}$$

$$= 0.75 \tag{8.12}$$

The change in refractive index (Δn) required for a bandgap shift is 0.75.

8.3.5 Imaging and Displays

The control over light propagation offered by photonic crystals contributes to advancements in imaging technologies and displays. Photonic crystal structures are utilized to create devices such as lenses and beam splitters, enabling the manipulation of light for improved image resolution and display quality.

8.3.6 Example: Photonic Crystal Lens

Consider a photonic crystal lens designed with a focal length (f) of 1 mm and a bandgap wavelength (λ) of 600 nm. The curvature radius (R) of the lens can be calculated using the lensmaker's formula:

$$\frac{1}{f} = (n - 1)\left(\frac{1}{R_1} - \frac{1}{R_2}\right) \tag{8.13}$$

where n is the refractive index.

$$f = 1\,\text{mm}$$

$$\lambda = 600\,\text{nm}$$

$$n = \frac{\lambda}{\text{Bandgap} + \lambda}$$

$$= \frac{600}{1400 + 600}$$

$$= 0.3$$

$$R = \frac{1}{(n-1)\left(\frac{1}{f} + \frac{1}{f}\right)}$$

$$= \frac{1}{(0.3 - 1)\left(\frac{1}{1 \times 10^{-3}} + \frac{1}{1 \times 10^{-3}}\right)}$$

$$\approx -1.67\,\text{m} \tag{8.14}$$

The curvature radius (R) of the lens is approximately $-1.67\,\text{m}$.

Chapter 9

Optical Signal Processing

9.1 Fourier Optics

Fourier Optics provides a mathematical description of how lenses and other optical elements transform the spatial distribution of light. The fundamental concept is based on the Fourier Transform, which allows us to express a complex optical field as a sum of simple sinusoidal components.

9.1.1 Fourier Transform

The Fourier Transform of a function $f(x)$ is defined as:

$$\mathcal{F}\{f(x)\} = F(u) = \int_{-\infty}^{\infty} f(x)\, e^{-i2\pi ux}\, dx \tag{9.1}$$

where u is the spatial frequency, i is the imaginary unit, and $\mathcal{F}$ denotes the Fourier Transform.

9.1.2 Example: Fourier Transform of Gaussian Function

Consider a Gaussian function given by:

$$f(x) = e^{-\alpha x^2} \tag{9.2}$$

The Fourier Transform of this Gaussian function is:

$$F(u) = \sqrt{\frac{\pi}{\alpha}}\, e^{-\pi^2 u^2/\alpha} \tag{9.3}$$

This example illustrates the relationship between a function in spatial domain and its frequency representation.

9.1.3 Fourier Optics in Imaging

In imaging systems, Fourier Optics plays a crucial role. The spatial frequency content of an image is directly related to the size and shape of the objects being imaged. The optical system, including lenses and apertures, affects the spatial frequencies that are transmitted and, consequently, the quality of the final image.

9.1.4 Example: Imaging System Transfer Function

The transfer function of an imaging system, denoted as $H(u, v)$, is the Fourier Transform of the Point Spread Function (PSF) $h(x, y)$. The PSF characterizes how a point source in the object space is spread in the image space. The transfer function is given by:

$$H(u, v) = \mathcal{F}\{h(x, y)\} \tag{9.4}$$

Understanding the transfer function allows engineers to analyze and design imaging systems for specific applications.

9.1.5 Fourier Optics in Signal Processing

Fourier Optics is widely used in signal processing applications. Optical systems can perform operations such as filtering, convolution, and correlation in the frequency domain, providing a unique approach to signal manipulation.

9.1.6 Example: Optical Convolution

The optical convolution of two functions $f(x, y)$ and $g(x, y)$ is given by the product of their Fourier Transforms:

$$f(x, y) * g(x, y) = \mathcal{F}^{-1}\{F(u, v) \cdot G(u, v)\} \tag{9.5}$$

This allows optical systems to perform convolution operations efficiently.

9.2 Applications of Fourier Optics

Fourier Optics finds applications in various fields, ranging from imaging and signal processing to holography and diffraction. The ability to analyze and manipulate optical signals in the frequency domain opens up avenues for innovative technologies.

9.2.1 Holography

Holography, a technique based on Fourier Optics, captures both the amplitude and phase information of an optical wavefront. The holographic plate records the interference pattern between the reference beam and the object beam, allowing the reconstruction of a three-dimensional image.

9.2.2 Example: Holographic Reconstruction

The holographic reconstruction of an object wave $O(x, y)$ from a hologram $H(x, y)$ is given by:

$$O(x, y) = \mathcal{F}^{-1}\{H(u, v) \cdot R(u, v)\} \tag{9.6}$$

where $R(u, v)$ is the complex conjugate of the reference wave.

9.2.3 Diffraction Gratings

Fourier Optics is essential in the design and analysis of diffraction gratings. Gratings consist of periodic structures that spatially modulate the intensity of diffracted light, allowing the separation and manipulation of different wavelength components.

9.2.4 Example: Grating Equation

The angular dispersion (θ_m) of the m-th diffraction order for a grating with a periodicity d is given by the grating equation:

$$\sin(\theta_m) = m\frac{\lambda}{d} \tag{9.7}$$

This equation illustrates how Fourier Optics principles are employed to understand the diffraction patterns produced by gratings.

9.3 Optical Filters

Optical filters are crucial elements in manipulating the spectral characteristics of light. They can be categorized into various types, including bandpass filters, low-pass filters, high-pass filters, and notch filters, each serving specific purposes in optical signal processing.

9.3.1 Bandpass Filter

A bandpass filter transmits a specific range of wavelengths while blocking others. The transmission function $T(\lambda)$ of a bandpass filter is defined as:

$$T(\lambda) = \begin{cases} 1 & \text{if } \lambda \text{ is within the passband} \\ 0 & \text{otherwise} \end{cases} \tag{9.8}$$

9.3.2 Example: Gaussian Bandpass Filter

Consider a Gaussian bandpass filter with a center wavelength (λ_0) of 800 nm and a bandwidth ($\Delta\lambda$) of 50 nm. The transmission function is given by:

$$T(\lambda) = e^{-\frac{(\lambda-\lambda_0)^2}{2\sigma^2}} \tag{9.9}$$

where σ is related to the bandwidth.

$$\lambda_0 = 800\,\text{nm}$$

$$\Delta\lambda = 50\,\text{nm}$$

$$\sigma = \frac{\Delta\lambda}{2\sqrt{2\ln(2)}}$$

$$\approx 21.2\,\text{nm}$$

$$T(\lambda) = e^{-\frac{(\lambda-800)^2}{2\times(21.2)^2}} \tag{9.10}$$

This example demonstrates the design of a Gaussian bandpass filter.

9.3.3 Low-Pass Filter

A low-pass filter transmits wavelengths below a certain cutoff value and attenuates higher wavelengths. The transmission function $T(\lambda)$ of a low-pass filter is defined as:

$$T(\lambda) = \begin{cases} 1 & \text{if } \lambda < \lambda_{\text{cutoff}} \\ 0 & \text{if } \lambda \geq \lambda_{\text{cutoff}} \end{cases} \tag{9.11}$$

9.3.4 Example: Rectangular Low-Pass Filter

Consider a rectangular low-pass filter with a cutoff wavelength (λ_{cutoff}) of 600 nm. The transmission function is given by:

$$T(\lambda) = \begin{cases} 1 & \text{if } \lambda < 600\,\text{nm} \\ 0 & \text{if } \lambda \geq 600\,\text{nm} \end{cases} \tag{9.12}$$

This example illustrates a simple rectangular low-pass filter.

9.3.5 High-Pass Filter

A high-pass filter transmits wavelengths above a certain cutoff value and attenuates lower wavelengths. The transmission function $T(\lambda)$ of a high-pass filter is defined as:

$$T(\lambda) = \begin{cases} 1 & \text{if } \lambda \geq \lambda_{\text{cutoff}} \\ 0 & \text{if } \lambda < \lambda_{\text{cutoff}} \end{cases} \tag{9.13}$$

9.3.6 Example: Triangular High-Pass Filter

Consider a triangular high-pass filter with a cutoff wavelength (λ_{cutoff}) of 700 nm. The transmission function is given by:

$$T(\lambda) = \begin{cases} \frac{\lambda - 700}{100} & \text{if } \lambda \geq 700\,\text{nm} \\ 0 & \text{if } \lambda < 700\,\text{nm} \end{cases} \tag{9.14}$$

This example demonstrates a triangular high-pass filter.

9.3.7 Notch Filter

A notch filter transmits most wavelengths but attenuates a specific narrow band of wavelengths. The transmission function $T(\lambda)$ of a notch filter is defined as:

$$T(\lambda) = \begin{cases} 0 & \text{if } \lambda \in [\lambda_1, \lambda_2] \\ 1 & \text{otherwise} \end{cases} \tag{9.15}$$

9.3.8 Example: Notch Filter for Sodium Doublet

Consider a notch filter designed to block the wavelengths of the sodium doublet, approximately 589 nm. The transmission function is given by:

$$T(\lambda) = \begin{cases} 0 & \text{if } \lambda \in [588.9, 589.1] \\ 1 & \text{otherwise} \end{cases} \tag{9.16}$$

This example illustrates the design of a notch filter for specific spectral lines.

9.4 Applications of Optical Filters

Optical filters find widespread applications in various areas, including telecommunications, spectroscopy, and imaging. Their ability to selectively manipulate the spectral content of light contributes to the advancement of optical technologies.

9.4.1 Telecommunications

In telecommunications, optical filters are employed to isolate specific wavelength channels in optical communication systems. Dense Wavelength Division Multiplexing (DWDM) systems utilize bandpass filters to separate and combine multiple optical signals over a single optical fiber.

9.4.2 Example: DWDM Channel Isolation

Consider a DWDM system with multiple channels spaced at $100\,\text{GHz}$ intervals. Bandpass filters are used to isolate individual channels. The transmission function for each channel can be represented as:

$$T(\lambda) = \begin{cases} 1 & \text{if } \lambda \in [\lambda_i - 0.05\,\text{nm}, \lambda_i + 0.05\,\text{nm}] \\ 0 & \text{otherwise} \end{cases} \tag{9.17}$$

where λ_i represents the center wavelength of each channel.

9.4.3 Spectroscopy

In spectroscopy, optical filters play a vital role in isolating specific spectral lines for analysis. Bandpass filters with precise wavelength characteristics enable researchers to study the absorption and emission spectra of various materials.

9.4.4 Example: Fluorescence Imaging

Fluorescence imaging often utilizes optical filters to isolate the emission wavelength of fluorophores. A bandpass filter with a transmission function centered around the fluorophore's emission peak enhances the signal-to-noise ratio in fluorescence microscopy.

9.4.5 Imaging

In imaging systems, optical filters contribute to improving image quality by selectively transmitting or blocking specific wavelengths. Color filters in digital cameras, for instance, help capture accurate colors by filtering incident light.

9.4.6 Example: RGB Color Filters

In digital imaging, RGB color filters are commonly used to capture red, green, and blue components separately. The transmission functions for each color channel are defined as:

$$T_{\text{red}}(\lambda) = \begin{cases} 1 & \text{if } \lambda \in [620\,\text{nm}, 750\,\text{nm}] \\ 0 & \text{otherwise} \end{cases} \tag{9.18}$$

$$T_{\text{green}}(\lambda) = \begin{cases} 1 & \text{if } \lambda \in [500\,\text{nm}, 620\,\text{nm}] \\ 0 & \text{otherwise} \end{cases} \tag{9.19}$$

$$T_{\text{blue}}(\lambda) = \begin{cases} 1 & \text{if } \lambda \in [400\,\text{nm}, 500\,\text{nm}] \\ 0 & \text{otherwise} \end{cases} \tag{9.20}$$

This example illustrates the design of RGB color filters.

Chapter 10

Optoelectronic Materials

In this chapter, we explore the fundamental aspect of Optoelectronic Materials, with a specific focus on Semiconductor Materials. Optoelectronic materials play a pivotal role in the development of devices that interact with light, such as photodetectors, light-emitting diodes (LEDs), and solar cells. Semiconductor materials, in particular, exhibit unique optoelectronic properties that are exploited in a wide range of applications.

10.1 Semiconductor Materials for Optoelectronics

Semiconductor materials are at the heart of optoelectronic devices, offering the ability to control the flow of charge carriers and interact with photons. The electronic band structure of semiconductors determines their optical and electrical properties, making them versatile in optoelectronic applications.

10.1.1 Band Structure of Semiconductors

The band structure of a semiconductor describes the distribution of energy levels for electrons in the material. It consists of a valence band and a conduction band, separated by a bandgap. The energy required to move an electron from the valence band to the conduction band is the bandgap energy (E_g).

$$E_\mathrm{g} = E_\mathrm{CB} - E_\mathrm{VB} \tag{10.1}$$

10.1.2 Example: Bandgap Energy of Silicon

For silicon, a widely used semiconductor, the bandgap energy is approximately 1.1 eV. This value determines the minimum energy of photons required to generate electron-hole pairs in silicon-based photodetectors.

10.1.3 Optical Absorption in Semiconductors

The absorption of light in semiconductors occurs when photons with energy greater than the bandgap energy are absorbed, creating electron-hole pairs. The absorption coefficient (α) is related to the optical intensity (I) and the incident power (P_in) as follows:

$$I = I_0 e^{-\alpha x} \tag{10.2}$$

where I_0 is the initial intensity, x is the distance traveled in the material, and α is the absorption coefficient.

10.1.4 Example: Optical Absorption in Gallium Arsenide (GaAs)

For GaAs, another common semiconductor with a bandgap energy of approximately 1.4 eV, the absorption coefficient can be calculated using the relation:

$$\alpha(\lambda) = \frac{A}{\lambda^n} \tag{10.3}$$

where A and n are material-dependent constants. This example demonstrates how to calculate the absorption coefficient for GaAs at a specific wavelength.

10.2 Applications of Semiconductor Materials

Semiconductor materials find extensive use in various optoelectronic applications, shaping the landscape of modern technology. The ability to control the movement of charge carriers and the interaction with light makes semiconductors indispensable in devices that enable communication, sensing, and energy harvesting.

10.2.1 Photodetectors

Photodetectors are devices that convert incident light into an electric current. Semiconductors with suitable bandgap energies are used to design photodetectors with specific spectral sensitivity. Silicon photodetectors, for instance, are sensitive to visible and near-infrared light.

10.2.2 Example: Photocurrent in Silicon Photodetector

The photocurrent (I_{photo}) generated in a silicon photodetector can be expressed as:

$$I_{\text{photo}} = q \cdot G \cdot P_{\text{in}} \tag{10.4}$$

where q is the charge of an electron, G is the photoconductive gain, and P_{in} is the incident optical power. This example illustrates the basic formula for calculating photocurrent in a silicon photodetector.

10.2.3 Light-Emitting Diodes (LEDs)

LEDs are semiconductor devices that emit light when current flows through them. The recombination of electron-hole pairs in the semiconductor material results in the emission of photons. Different semiconductor materials produce LEDs with different colors.

10.2.4   Example: Wavelength of LED Emission

The wavelength (λ) of light emitted by an LED is related to the bandgap energy (E_g) by the energy-wavelength relation:

$$\lambda = \frac{c}{f} = \frac{hc}{E_\mathrm{g}} \tag{10.5}$$

where c is the speed of light, h is Planck's constant, and f is the frequency of emitted light. This example demonstrates how to calculate the wavelength of emitted light based on the bandgap energy.

10.2.5   Solar Cells

Semiconductor materials are crucial in solar cells for converting sunlight into electrical power. Photons with energy greater than the bandgap energy create electron-hole pairs, generating a photocurrent that can be harvested as electrical power.

10.2.6   Example: Efficiency of Solar Cell

The efficiency (η) of a solar cell is determined by the ratio of the maximum power point (P_max) to the incident power (P_in):

$$\eta = \frac{P_\mathrm{max}}{P_\mathrm{in}} \tag{10.6}$$

This example illustrates the calculation of the efficiency of a solar cell.

10.3   Organic Optoelectronic Materials

Organic optoelectronic materials have gained significant attention due to their flexibility, light weight, and potential for low-cost manufacturing processes. These materials are based on organic compounds, often polymers or small molecules, and find applications in devices such as organic light-emitting diodes (OLEDs), organic solar cells, and organic field-effect transistors.

10.3.1 Conjugated Polymers

Conjugated polymers are a class of organic materials with alternating single and double bonds along the polymer chain. This conjugation leads to enhanced electronic properties, making them suitable for optoelectronic applications. The electronic structure of a conjugated polymer can be represented as:

$$\text{Polymer Chain:} \quad -C = C - C = C-$$
$$\text{Electronic Structure:} \quad \text{HOMO} \quad \pi\text{-orbitals} \quad \text{LUMO} \tag{10.7}$$

10.3.2 Example: Bandgap of Conjugated Polymer

The bandgap (E_g) of a conjugated polymer can be estimated using the empirical equation:

$$E_g \approx \frac{2.5}{L} \tag{10.8}$$

where L is the conjugation length in angstroms. This example demonstrates a simple formula to estimate the bandgap of a conjugated polymer based on its conjugation length.

10.3.3 Organic Light-Emitting Diodes (OLEDs)

OLEDs are optoelectronic devices that emit light when an electric current is applied. They consist of organic layers sandwiched between electrodes. The emission color is determined by the organic materials used. The general working principle involves electron-hole recombination and subsequent light emission.

10.3.4 Example: Emission Wavelength in OLEDs

The emission wavelength (λ) in OLEDs is related to the energy gap between the highest occupied molecular orbital (HOMO) and the lowest unoccupied molecular orbital (LUMO) of the organic material:

$$\lambda = \frac{c}{E_{\text{HOMO-LUMO}}} \tag{10.9}$$

where c is the speed of light. This example illustrates how to calculate the emission wavelength based on the HOMO-LUMO energy gap.

10.3.5   Organic Solar Cells

Organic solar cells harness the photovoltaic effect in organic materials to convert sunlight into electricity. The active layer typically consists of a blend of electron-donor and electron-acceptor organic compounds. The power conversion efficiency (η) of a solar cell can be calculated using the equation:

$$\eta = \frac{V_{\text{oc}} \cdot J_{\text{sc}} \cdot FF}{P_{\text{in}}} \tag{10.10}$$

where V_{oc} is the open-circuit voltage, J_{sc} is the short-circuit current density, FF is the fill factor, and P_{in} is the incident power.

10.3.6   Example: Efficiency of Organic Solar Cell

Consider an organic solar cell with $V_{\text{oc}} = 0.6\,\text{V}$, $J_{\text{sc}} = 20\,\text{mA/cm}^2$, $FF = 0.7$, and $P_{\text{in}} = 100\,\text{mW/cm}^2$. The efficiency can be calculated using the provided formula.

Chapter 11

Photovoltaic Devices

In this chapter, we delve into the realm of Photovoltaic Devices, with a particular focus on Solar Cells. Solar cells are devices that convert sunlight into electrical energy through the photovoltaic effect. Understanding the principles and characteristics of solar cells is crucial for harnessing solar energy for various applications.

11.1 Solar Cells

Solar cells, also known as photovoltaic cells, are semiconductor devices that generate electrical power when exposed to sunlight. The most common material used in solar cells is silicon. The operation of a solar cell involves the absorption of photons, the creation of electron-hole pairs, and the flow of these charge carriers to generate an electric current.

11.1.1 Photovoltaic Effect

The photovoltaic effect in a solar cell is the phenomenon where photons with energy greater than the bandgap of the semiconductor material create electron-hole pairs. The bandgap energy (E_g) is a crucial parameter that determines the minimum energy of photons needed for this process.

$$E_g = E_{\mathrm{CB}} - E_{\mathrm{VB}} \tag{11.1}$$

11.1.2 Example: Silicon Solar Cell Bandgap

For silicon solar cells, which are widely used, the bandgap energy is approximately 1.1 eV. This means that photons with energy greater than 1.1 eV can contribute to the generation of electron-hole pairs.

11.1.3 Solar Cell Efficiency

The efficiency (η) of a solar cell is a critical parameter that quantifies its ability to convert sunlight into electrical power. The efficiency is determined by the ratio of the maximum power point (P_{max}) to the incident power (P_{in}).

$$\eta = \frac{P_{\mathrm{max}}}{P_{\mathrm{in}}} \tag{11.2}$$

11.1.4 Example: Calculating Solar Cell Efficiency

Consider a silicon solar cell with an open-circuit voltage (V_{oc}) of 0.6 V, a short-circuit current density (J_{sc}) of 30 mA/cm^2, and a fill factor (FF) of 0.8. The incident sunlight power (P_{in}) is 100 mW/cm^2. The efficiency can be calculated using the provided formula.

11.1.5 Solar Cell Characteristics

Solar cells can be characterized by their current-voltage ($I - V$) curve, which depicts the relationship between the current flowing through the cell and the voltage across its terminals. The $I - V$ curve is essential for analyzing the performance of a solar cell under different operating conditions.

11.1.6 Example: Analyzing $I - V$ Characteristics

Analyze the $I - V$ characteristics of a solar cell with the following parameters: $V_{oc} = 0.7\,\text{V}$, $J_{sc} = 25\,\text{mA/cm}^2$, $FF = 0.75$, and $P_{in} = 120\,\text{mW/cm}^2$. Plot the $I - V$ curve and discuss the key points.

11.2 Photovoltaic Materials

Photovoltaic materials are semiconductors that form the active layers of solar cells, facilitating the conversion of sunlight into electricity. These materials possess unique electronic and optical properties that enable efficient absorption and generation of charge carriers.

11.2.1 Semiconductor Bandgap

The bandgap of a semiconductor is a key parameter that determines its ability to absorb photons and generate electron-hole pairs. The bandgap energy (E_g) is the energy difference between the valence band (E_{VB}) and the conduction band (E_{CB}).

$$E_g = E_{CB} - E_{VB} \tag{11.3}$$

11.2.2 Example: Bandgap of Silicon

Silicon is a widely used semiconductor material in solar cells, and it has a bandgap energy of approximately $1.1\,\text{eV}$. This value influences the range of wavelengths of light that can be absorbed by silicon.

11.2.3 Absorption Coefficient

The absorption coefficient (α) quantifies the rate at which a material absorbs light as a function of the wavelength (λ). It is a crucial parameter in determining the material's ability to absorb sunlight.

$$I(x) = I_0 \cdot e^{-\alpha x} \tag{11.4}$$

where $I(x)$ is the intensity of light at a depth x in the material, I_0 is the initial intensity, and α is the absorption coefficient.

11.2.4 Example: Calculating Absorption Depth

For a given material with an absorption coefficient of $\alpha = 10\,\text{cm}^{-1}$, calculate the absorption depth (x) for sunlight with a wavelength of $\lambda = 500\,\text{nm}$.

11.2.5 Carrier Generation Rate

The rate of carrier generation (G) in a photovoltaic material is influenced by the incident sunlight intensity (I_{sun}) and the absorption coefficient (α).

$$G = \alpha \cdot I_{\text{sun}} \tag{11.5}$$

11.2.6 Example: Carrier Generation in Silicon

For a silicon solar cell with an incident sunlight intensity of $100\,\text{mW/cm}^2$ and an absorption coefficient of $\alpha = 10\,\text{cm}^{-1}$, calculate the carrier generation rate.

Appendix

In this appendix, we provide additional information and resources related to the formulas and concepts covered in the main chapters of the book "Optoelectronics: A Formula Handbook." This section serves as a reference guide and includes supplementary materials to enhance the reader's understanding of optoelectronic principles.

11.3   Useful Constants

$$\text{Speed of Light } (c): \qquad c = 3.00 \times 10^8 \,\text{m/s} \qquad (11.6)$$

$$\text{Planck's Constant } (h): \qquad h = 6.63 \times 10^{-34} \,\text{J} \cdot \text{s} \qquad (11.7)$$

$$\text{Elementary Charge } (e): \qquad e = 1.60 \times 10^{-19} \,\text{C} \qquad (11.8)$$

$$\text{Boltzmann's Constant } (k): \qquad k = 1.38 \times 10^{-23} \,\text{J/K} \qquad (11.9)$$

These constants are frequently used in optoelectronic calculations and provide fundamental values for understanding the behavior of light and matter.

11.4   Quantum Circuit for Quantum Optics

Quantum optics often involves the representation of quantum circuits to describe the interactions of photons and quantum systems. The following quantum circuit illustrates a basic quantum optics setup for a beamsplitter:

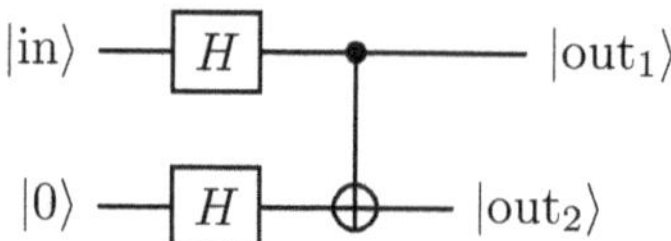

This quantum circuit represents a beamsplitter that transforms the input state $|\text{in}\rangle$ and an initial vacuum state $|0\rangle$ into two output states $|\text{out}_1\rangle$ and $|\text{out}_2\rangle$.

11.5 Fresnel Equations for Reflection and Transmission

The Fresnel equations describe the reflection and transmission coefficients of light at an interface between two media with different refractive indices. The equations for the amplitude reflection coefficient (r) and amplitude transmission coefficient (t) are given by:

$$r = \frac{n_1 - n_2}{n_1 + n_2} \tag{11.10}$$

$$t = \frac{2n_1}{n_1 + n_2} \tag{11.11}$$

Here, n_1 and n_2 are the refractive indices of the two media.

11.6 Example: Calculating Reflection and Transmission

Consider light incident on a glass-air interface ($n_1 = 1.5$, $n_2 = 1.0$). Calculate the amplitude reflection and transmission coefficients using the Fresnel equations.

$$r = \frac{1.5 - 1.0}{1.5 + 1.0} \tag{11.12}$$

$$t = \frac{2 \times 1.5}{1.5 + 1.0} \tag{11.13}$$

This example demonstrates the application of the Fresnel equations in determining the behavior of light at an interface.

Bibliography

Books

1. Saleh, B. E. A., & Teich, M. C. *Fundamentals of Photonics*. Wiley, 2007.

2. Yariv, A., & Yeh, P. *Optical Waves in Crystals: Propagation and Control of Laser Radiation*. Wiley, 2003.

3. Sze, S. M. *Physics of Semiconductor Devices*. Wiley, 1981.

4. Ghatak, A., & Thyagarajan, K. *Introduction to Fiber Optics*. Cambridge University Press, 1998.

Journal Articles

1. Agrawal, G. P. "Fiber-Optic Communication Systems." *Johns Hopkins APL Technical Digest*, vol. 16, no. 3, 1995, pp. 342-348.

2. Miller, D. A. B. "Device Requirements for Optical Interconnects to Silicon Chips." *Proceedings of the IEEE*, vol. 97, no. 7, 2009, pp. 1166-1185.

3. Almeida, V. R., Barrios, C. A., Panepucci, R. R., & Lipson, M. "All-Optical Control of Light on a Silicon Chip." *Nature*, vol. 431, no. 7012, 2004, pp. 1081-1084.

Conference Papers

1. Jalali, B., & Fathpour, S. "Silicon Photonics." In *IEEE Journal of Selected Topics in Quantum Electronics*, vol. 12, no. 6, 2006, pp. 1678-1687.

2. Birowosuto, M. D., Driessen, A., Duis, J., Leinse, A., & Roeloffzen, C. G. H. "Monolithic Integration of an InP-Based Photodetector for a Silicon Photonics Receiver." In *Optical Fiber Communication Conference*, 2012, paper OTh1H.4.

Online Resources

1. The Optical Society. "Optics and Photonics News." Available at: `https://www.osapublishing.org/opn`.

2. SPIE Digital Library. "Advances in Optoelectronics." Available at: `https://www.spiedigitallibrary.org/journals/advances-in-optoelectronics`.